STRATEGIC MARKETING FOR ELECTRIC UTILITIES

STRATEGIC MARKETING FOR ELECTRIC UTILITIES

CLARK W. GELLINGS

DILIP LIMAYE

Published by

THE FAIRMONT PRESS, INC.

700 Indian Trail

Lilburn, GA 30247

Library of Congress Cataloging-in-Publication Data

Gellings, Clark W.
Strategic marketing for electric utilities.

Includes index.
1. Electric power--Marketing. 2. Electric utilities.
I. Limaye, Dilip R. II. Title.
HD9685.A2G454 1988 363.6'2'0688 86-46138
ISBN 0-88173-037-8

Strategic Marketing for Electric Utilities

Published by The Fairmont Press, Inc.
700 Indian Trail
Lilburn, GA 30247

ISBN 0-88173-037-8 FP

ISBN 0-13-851585-9 PH

Printed in the United States of America

Distributed by Prentice Hall
A division of Simon & Schuster
Englewood Cliffs, NJ 07632

Prentice-Hall International (UK) Limited, London
Prentice-Hall of Australia Pty. Limited, Sydney
Prentice-Hall Canada Inc., Toronto
Prentice-Hall Hispanoamericana, S.A., Mexico
Prentice-Hall of India Private Limited, New Delhi
Prentice-Hall of Japan, Inc., Tokyo
Simon & Schuster Asia Pte. Ltd., Singapore
Editora Prentice-Hall do Brasil, Ltda., Rio de Janeiro

Strategic Marketing for Electric Utilities

Dilip R. Limaye & Clark W. Gellings

CONTENTS

Contributors

Laurel Andrews -- Synergic Resources Corporation, Seattle, WA
John H. Chamberlin -- Barakat, Howard & Chamberlin, Inc., Berkeley, CA
Todd D. Davis -- Synergic Resources Corporation, Bala Cynwyd, PA
Nancyann Emanuelson -- Price Waterhouse, New York, NY
Ahmad Faruqui -- Battelle-Columbus Division, Palo Alto, CA
John E. Flory -- Utility-Customer Interface, Inc., Davis, CA
Dennis Horgan -- Price Waterhouse, New York, NY
Larry E. Lewis -- Electric Power Research Institute, Palo Alto, CA
Nancy Licht -- Barakat, Howard & Chamberlin, Inc., Berkeley, CA
Craig McDonald -- Synergic Resources Corporation, Bala Cynwyd, PA
Louise M. Morman -- Price Waterhouse, New York, NY
Carol Sabo -- New York State Electric & Gas Corp., Binghamton, NY
Richard Tempchin -- Synergic Resources Corporation, Bala Cynwyd, PA

CHAPTER 1

Introduction

Clark W. Gellings and Dilip R. Limaye

CHANGES IN THE ELECTRICITY MARKETS

The first utility marketing program was undertaken in 1882 by the Edison Pearl Street New York Utility. The goal of this program was to increase daytime load; a fledgling technology -- the electric motor -- was chosen as the means to increase load during the off-peak (daytime) period. The target of the marketing program: industrial firms using less efficient technology.

Through the early decades of this century, the industry focused its marketing efforts on time-of-use, off-peak, and water heating control rates. Through the 1950s and '60s, a time of unprecedented economic expansion, the emphasis was on growth. Changes in the marketplace, resulting from the "energy crisis," led many utilities to abandon marketing entirely in the late '60s and early '70s. Others shifted their attention from marketing to conservation. By the mid-1970s, load management was in vogue and, coupled with increasing interest in conservation, led to renewed interest in marketing as a means to increase customer acceptance of utility programs and initiatives.

Today the traditional structure of the "energy market" is changing and will continue to change. As significant as the evolution of the fundamental structural changes in the electricity industry is the fundamental change in the basic business relationship between utility and customer that many believe is the catalyst for these structural changes. Changes to the relationships between electric utilities and their customers are accelerating the evolution of the industry from the traditional "end-to-end" regulated monopoly structure, where the price of electricity is set by regulation to meet a revenue requirement, to a competitive environment where the price is set by market value. The utility planner and manager must know how the individual customer and the market in aggregate determine value, what the components of value are, and what tools and techniques are needed to incorporate value into supply- and demand-side planning.

The basic business relationship includes four exchanges: provision of a product and/or service; payment by the buyer to the seller; information from the seller to the buyer (e.g., marketing and advertising); and information from the buyer to the seller (e.g., satisfaction or dissatisfaction), which indicates the likelihood of continued purchase of the product or service. It is not that this basic business relationship no longer functions within the electric utility industry or that the same four exchanges do not continue to exist; rather, each party's perspective toward these exchanges and their interrelationships has evolved over the last 15 years.

VALUE

Inherent in all business exchanges is a value-to-cost ratio. The buyer must perceive that the value of the product or service purchased is greater than the price or cost. If the value is below the cost he will refrain from further purchases or will look for substitutes that provide more value, or cost less and deliver the same value. The seller in turn must charge a price in excess of production cost. If, due to the nature of the market, the seller cannot charge a price sufficient to cover his production cost and return a profit, he will either go out of business or produce a different product that will enable him to succeed.

In the regulated monopoly structure, the utility is allowed to charge a price (rate) which reflects the cost of production plus a rate of return on capital investment sufficient to attract additional capital. The customer's perceived value has generally been thought to be based solely on a need for a highly reliable product delivered on demand. As long as the utility provided highly reliable electricity in the quantities demanded, it was believed that the customer would perceive a high value-to-cost ratio. However, the customer's perceived value of electricity and energy services is influenced by several other components in addition to reliability and cost. Traditionally, utilities have concentrated on reducing cost as the way to improve the ratio. However, another way is to increase the value. The utility industry must learn how best to produce a high value-to-cost ratio for all customers so that the best interests of the utility and the customer can be met within a more competitive business climate.

ELECTRICITY VALUE AND COST: TRENDS AND PROSPECTS

The customer's value-to-cost ratio has changed several times over the years. Following is a brief summary of the evolution of this ratio.

Pre-1973: Customer as Meter

The pre-1973 period has been called the "golden age" of utilities because value was very high in comparison to cost. Both components of the ratio were moving in positive directions. Value was moving up as more efficient electrical appliances entered the market, making electricity a more valuable commodity and service; and cost was going down as economies of scale and the construction of large central plants lowered the unit price for the industry. The industry did not have to market its product in the traditional sense; they could simply sell it. Demand growth was steady at more than 7% per year, and forecasts of demand growth for the 1970s and 1980s presumed a continuation of that pattern. Demand for electricity was tracking GNP growth and there was no indication that any circumstances would alter this relationship.

Marketing concerns of the pre-1973 period were focused on improving the measurement of usage at the meter and keeping a clear demarcation between the customer and the utility side of the meter.

Economic variables and their relationships to demand were the focal point of most research. Refinement of econometric models that depicted a more accurate relationship between demand and economic variables was the focus of many marketing programs because this relationship was believed to be immutable and was seen as the driving force in forecasting the need for new plant. At that time, it was believed that a steady 7% annual growth could be sustained for the rest of the century, and planning decisions were made accordingly. The customer's role as an actor in the planning process was essentially ignored. There was no real marketing nor any attempt to change consumption patterns other than to make access to electricity and electrical appliances easier by offering line extension credits to builders and promoting all-electric homes to increase market share.

1973-1980: Customer as Appliance

In this period, fuel costs, a major component of utility operating cost, rose sharply, creating an imbalance in the customer's value-to-cost ratio. Value can be measured in several ways, and during this period customers tended to measure it relative to the cost they had historically paid for electricity, even though the quality of service had not diminished. As their costs rose, their perception of value decreased. The customers, however, had few alternatives by which to realign the ratio other than reducing consumption. The utilities were under regulatory pressure to hold down their rate and fuel adjustment increases and were expected to provide conservation and load management services that would enable the customer to take steps to hold down costs. The focus of utility activity was to realign the value-to-cost ratio by reducing cost. Attempts at cost reduction included strategic conservation and load management programs on the demand side, and the refinement of forecasting and resource planning methods to reduce production costs on the supply side.

The initial focus of many utility marketing support programs in this period was on the appliances (end uses) and the efficiency levels that could be obtained through conservation, load management, and efficiency standards programs. The customer was assumed to be motivated solely by economics. Since participation was anticipated to be high -- the reason people would or would not participate was whether they had access to information or services, not whether they found value in saving -- the primary research problem was refining the engineering calculations used to measure and predict savings for incorporation into utility demand forecasts and resource plans.

It soon became clear to utilities and regulators alike, however, that the anticipated participation rates were not materializing and that, consequently, the forecasted savings would be in error. The first programs to be widely examined were time-of-use rates and direct load control programs. These were important programs to examine because they presented clear examples of the customer's value-to-cost ratio. Utilities were asking the customer to alter consumption patterns in a way that very often produced some degree of inconvenience or discomfort, raising a fundamental question: How much is inconvenience or discomfort worth?

The conventional wisdom held that a monetary incentive would suffice as a trade-off for inconvenience or discomfort, and that larger incentives would result in higher participation rates. If these assumptions were true, then the utility should be able to predict participation rates based on the size of the incentive. Utilities were willing to offer a fairly high incentive payment, but the results were not as encouraging as had been anticipated. Furthermore, regulators who were authorizing these incentive payments were beginning to ask whether the incentives were too high compared to the value of the load shift. This began to bring into question whether there was indeed a correlation between incentive size and participation.

A second fundamental question raised by the unexpectedly low participation rates was whether participation could be affected by the type of promotion or type of contact the utility used to market its programs. Did it matter whether the program was vigorously marketed? Were some marketing techniques better than others? How do marketing techniques interact with incentive levels and other program attributes?

Results of early research into participation rates in direct load control programs indicated that the degree of marketing support for the programs was a significant factor in improving participation. The median participation rate was some seven times higher for direct load control programs supported by intensive marketing than for programs with minimal marketing support. Intensive marketing was defined as direct customer contact, the most costly of marketing techniques. It was an important step to prove that intensive marketing made a difference, but the results begged the question of whether the cost of the marketing was justified in terms of the improved participation rate and the value of the actual savings or load shifts accrued.

As for the incentive question, results at this point were less clear. Research showed that incentives of some size were needed to induce participation, and that they were more crucial to sustaining continued participation than to gaining initial participation. But there did not seem to be any proof that there was a one-to-one correlation between incentive amount and participation such that the utility could increase the incentives to attain higher participation rates.

What the research did lead to was insight into the patterns of customer decisions. The customer apparently used different value-to-cost ratios in evaluating initial participation versus continuing participation. Also, the research indicated that arriving at a decision to participate in such programs was a dynamic process comprised of several stages. If the utility wished to influence the buying process, then it had to understand the components of the process and the factors that moved the customer through it.

The results of the early research into the customer's participation decision led the industry to examine the customer's buying process more carefully to see whether customer behavior could be predicted given different types of program offerings and incentive levels. If different factors affected the buying process, could those factors be adjusted to affect specific participation levels in a predictable way?

The first breakthrough in predictive models resulted from customer acceptance work done within the rate design study sponsored by the Electric Power Research Institute. This research revealed that there was indeed a "customer purchase paradigm" and that this paradigm could be used to quantify and predict the different responses produced by different features of a marketing program.

The approach used in this early work was based on an existing market model. This model suggests that beliefs about and feelings toward needs and alternative ways to fulfill those needs work together to form intentions. It is the intention which ultimately leads to behavior, i.e., the purchase of a product or service. This model offers a visual description of the idea that "nothing happens until someone buys something." And no one buys something without first forming an intention to do so.

This is a dynamic paradigm. Once a purchase is made, as the fulfillment of an intention, the experience/satisfaction resulting from that purchase creates new beliefs about and feelings toward the need and the solutions chosen. If the customer is satisfied with the product or service, he will have a positive feeling and a strong belief that the product or service should be repurchased. Conversely, dissatisfaction will lead to a search for an alternative solution to the need. In terms of utility demand-side programs, the customer's beliefs about and feelings toward the utility and the products or services offered will influence his intention to participate in such programs.

This early research was very important as a first step toward developing a predictive framework or model that would anticipate customer participation and integrate customer responses into the planning structure of the utility -- but it was only a first step. What it was able to do was to establish that there was a weak link between economic incentives and participation rates and, most importantly, that no simple model could predict participation rates using a program attribute as a variable. This research also showed that the level and type of marketing effort did have an impact on participation rates, that the customer was sensitive to more than economics, and that the impacts to the utility's system load could not be forecasted solely by using engineering calculations of appliance efficiency.

What this early effort could not do was predict relationships between buying behavior and customer attitudes toward specific energy services and demand-side options. There also was an insufficient data base to indicate how these new insights into customer buying behavior could be integrated into the utility planning process. Although much had been learned, these lessons only expanded the number of questions that required further research.

1980-1986: Customer as Economic Man

With the maturation of strategic conservation and load management programs, significant third-party power development, decreasing oil and gas prices, and increasing self-generation, the value-to-cost ratio became more segmented across and within customer classes.

Commercial customers have perceived the highest value-to-cost ratio, as evidenced by the small amount of demand-side/cost-control activity in this sector. The cost of electricity and energy services to commercial customers is still a relatively small component of their cost of doing business, and the value of their utility service is still higher than the cost.

Industrial customers have been shown to perceive the lowest value-to-cost ratio when evaluating their utility service. The industrial class has in many instances borne a disproportionate fixed cost burden, and they are the customers moving rapidly to alternative energy sources and self-generation.

Residential customers seem to perceive a two-tiered value-to-cost ratio. This interesting fact is found in an Edison Electric Institute (EEI) nationwide survey conducted by Cambridge Reports, which indicated that customers look at their electric utility in two ways: as a product provider and as a service provider. When the utility is viewed as a product provider, the customer perceives a high value-to-cost ratio; however, when it is viewed as a service provider, the customer perceives a lower value-to-cost ratio. The survey also indicated that the two ratios are correlated in such a way that the utility can trade off better service in exchange for higher rates without disturbing the ratio.

As the industry moved into the mid-1980s and utility demand-side programs matured, it became even more important to flesh out the early developments in customer research so that utility marketing efforts could be as cost-effective and predictable as possible. What was needed was more formalization of the demand-side options available, the methodology for selecting programs, and the techniques for predicting results.

These industry needs led to the development of a formalized decision framework by the Electric Power Research Institute (EPRI) and EEI. This framework was to provide predictive models of demand-side program acceptance and results, and to clearly define the role of utility marketing as the method for ensuring program success.

The principal early product of the EPRI/EEI work is the five-volume Guidebook for Demand-Side Management, which outlines basic frameworks for decision-making in marketing within the framework of overall utility planning, and catalogs known acceptance information. The Guidebook presented the six generic categories of utility market implementation methods -- customer education, direct customer contact, trade ally cooperation, advertising and promotion, alternative pricing, and direct incentives -- that could be applied to any of the over 150 demand-side program options also identified in the Guidebook.

These tools are the fundamental building blocks for the predictive model. A utility would start with one of the six generic load shape objectives -- strategic conservation, peak clipping, peak shifting, valley filling, strategic load growth, and flexible load growth -- check the catalog of demand-side techniques for the ones most likely to produce the desired load shape, and then match the techniques with the most suitable marketing techniques to ensure the desired participation rate. The end product is a system for making demand-side planning decisions that for the first time puts together the tools for predicting program results, a system which explicitly ties load shape objectives to marketing techniques.

Although the demand-side management (DSM) framework is a crucial development in the evolution of utility marketing to a more customer-driven approach, it still is based on a value-to-cost ratio rooted in the utility's load shape objectives. The DSM objective of providing a "road map" to economic benefits by optimizing load shape presumes that the customers will benefit equally from such a change, and indeed will value that change as highly as the utility. It is obvious that customers would like rate stability -- or better yet, lower rates -- and that load shape optimization can help to bring that about. However, just because utility and customer each benefit from an improved load shape, it does not necessarily follow that they will benefit equally, or that the customer's acceptance and response to DSM programs will be motivated solely by economics.

Indeed, the electric utility industry's experience with the effects of the different marketing techniques used to achieve load shape objectives revealed that different marketing tools were more effective for different customers and for different stages of their buying processes. This experience confirmed the existence of a definite customer buying process, and fleshed out the notion of "buyer readiness" -- that there are different factors involved in going from intention to action and from initial action to continued action (satisfaction).

It also became clear that different customers were at different stages of buyer readiness and that different customers would be motivated to move through the buying stages for different reasons. Market segments had to be developed by defining common buying behavior patterns which would help to identify common stages of buyer readiness, and more importantly, common programs and marketing techniques that would move them along the buying process.

1987 and Beyond: Customer as Complex Consumer

Two important developments took place in the mid-'80s that made it easier for utilities to move from a selling orientation in which the utility's objectives drive the planning to a marketing orientation in which the customer's needs drive the planning: a reevaluation of the appropriateness of the marketing function itself (marketing having been totally out of favor since the early 1970s); and improvements in quantification techniques in the overall discipline of market research, which allowed for the measurement of attitudes and preferences which previously had been too subjective for use in utility planning.

Utility marketing departments of the 1950s and 1960s were abandoned in the 1970s and early 1980s; in many cases, the marketing specialists were told to focus on promoting conservation. The utilities were asked to adapt from selling electricity to selling "savings." Now, utilities in many parts of the country have surplus power, and there is growing competition from other fuels and self-generation. Utilities are beginning to reestablish their marketing departments, and the regulatory community is slowly beginning to accept the concept. As this trend continues, it will become more and more important for utilities to have a well-defined and well-developed marketing function complete with the latest techniques for understanding the customer's energy service needs and preferences--and with the ability to design programs to meet those needs.

The market research techniques now available to the utility industry have evolved since World War II into a set of sophisticated tools that permit analyses of complex behaviors. Several factors in this development are important to note. First, the U.S. economy has evolved from a production-oriented economy to a service economy. Marketing research has had to change its focus to the customer's needs for and attitudes toward services, which often are less tangible than those for production goods. Second, in the last 30 years, there have been significant increases in the numbers and types of new services available to the customer. Third, there have been significant changes in and expansion of the media available to advertisers, making it much more important to understand media preferences. Fourth, there have been significant demographic changes in the structure of the "typical" American family; there are many more segments to understand. Fifth, advances in computer technology and the introduction of the Universal Product Code system have allowed for exponential growth in the ability to process information and provide statistical analysis that had previously been prohibitively expensive. And finally, developments in mathematical psychology techniques, which have benefited from the computer revolution, have produced more sophisticated measurement techniques, allowing quantification of attitudes and preferences.

With these developments, the utility industry is now in a position to take the groundbreaking work from the DSM project and other early EPRI marketing support programs and move into a refined customer-driven approach to utility planning. In order to do this, a cleaner set of predictive models which explain the customer's buying process is needed; these models must also allow for accurate market segmentation. These "choice" models will take advantage of marketing support research into the customer's buying stages, buyer readiness, barriers to movement through the stages, attitudes toward and awareness of different advertising techniques, market segmentation, and the lessons learned by utilities that have implemented demand-side programs using the framework. What is needed is a synthesizing effort that will produce predictive models particular to the utility industry, models which use the latest market research techniques and which are grounded in the crucial understanding that the consumer of energy services is a complex consumer.

EPRI's Customer Preference and Behavior project, which will feature a closely coordinated national survey of utility customers that will collect and analyze data on a variety of behavioral parameters, is just such an effort. These data will be used to develop a series of computer models that simulate customer responses to DSM program marketing and operations. Case studies, consisting of DSM test programs, will then be performed at a number of different utilities in order to gather information on actual customer responses to the program parameters (e.g., promotional efforts, incentives, DSM strategies). The case study findings will be used to calibrate and refine the computer models so that they can be a reliable predictive tool of customer acceptance and response to DSM programs.

A GUIDE TO THIS BOOK

This book is designed to familiarize the utility practitioner with the concepts of strategic marketing as well as to provide guidance on the methods and techniques used to implement marketing strategies and programs. The next three chapters focus on developing an understanding of the customers' perspectives, needs, preferences, and behavior -- the first critical step in strategic marketing. Based on research sponsored by EPRI, these chapters, authored by Gellings, Flory, and Lewis, provide the utility planner with an excellent starting point for the design of marketing efforts. Chapter 5 by Limaye addresses in more detail the issues related to customer decision-making in the business environment, a process in which many individuals participate in and influence the final decision.

Chapters 6 through 8 delve into the mechanics of establishing successful marketing programs, addressing methods and techniques for market segmentation, test marketing, and market penetration assessment. Chapter 9 discusses the new competitive challenges faced by electric utilities and the potential responses to these challenges. Chapter 10 tackles an important practical issue faced by most utilities -- how to convert from the conservation orientation of the last decade to the marketing focus needed in the next decade.

Chapters 11 through 14 provide specific examples and case studies of utility marketing programs. Reading about the successes -- and failures -- of such programs is likely to be as instructive as understanding the underlying concepts and theories. These chapters document some of the lessons learned by many utilities -- investor-owned, publicly owned, and rural electric cooperatives -- from the implementation of specific programs.

The final two chapters are a little more provocative, perhaps even somewhat controversial. They are intended to stimulate the reader to think about the future directions of utility marketing. In Chapter 15, Faruqui provides interesting analogies between marketing strategies and warfare strategies; in Chapter 16, Davis urges utilities to move beyond the marketing myopia that has characterized past utility efforts. While the reader may not agree with all of the assertions in these final two chapters, it is hoped that they will provide new perspectives on utility marketing, and encourage innovation and creativity in future utility marketing efforts.

CHAPTER 2

The Customer's Perspective -- Is It the Same as the Utility's?

Clark W. Gellings and John E. Flory

INTRODUCTION

Traditionally, marketing has been viewed as the process whereby potential customers are encouraged to buy, and current customers are encouraged to buy more. For electric utilities, this rule of thumb translated into the generalized promotion of any new or additional electrical energy use. Later, increasing rates caused by rising production costs led to lowered energy consumption which, in spiral fashion, led to further increases in rates, decreases in consumption, and steadily declining utility earnings. Consequently, electric utilities and their regulators were prompted to reexamine the marketing process.

This reexamination has led to a new marketing approach that is carefully planned and strategically targeted. The approach is based on the belief that supply and demand function dynamically, each influencing the other. The new marketing approach has sought to tailor electric products/services to more closely match both customers' needs and utilities' characteristics.

A large part of this innovative approach is demand-side management (DSM). DSM is a sophisticated planning/decision framework that assists utility planners in deciding the subject and style of market promotion.

Use of DSM benefits both utilities and their customers. For utilities, DSM augments the range of alternatives for actively influencing electricity demand, and thus produces desired changes in the utilities' load shapes (Figure 2-1). Demand-side management includes conservation, strategic load growth, and load management--both direct control and voluntary (hard and soft). It also gives the utility the flexibility to successfully meet the challenges of an often changing marketplace. On the other hand, DSM provides electric consumers with the much-desired opportunity to participate in better control and management of their total energy cost and usage. Over 300 of the 3,000 or so U.S. electric utilities have already embraced some form of demand-side management. A recent survey by the Electric Power Research Institute (EPRI) identified 1,650 separate DSM programs.

Figure 2-1. USE OF DSM CAN ALTER UTILITY LOAD SHAPES

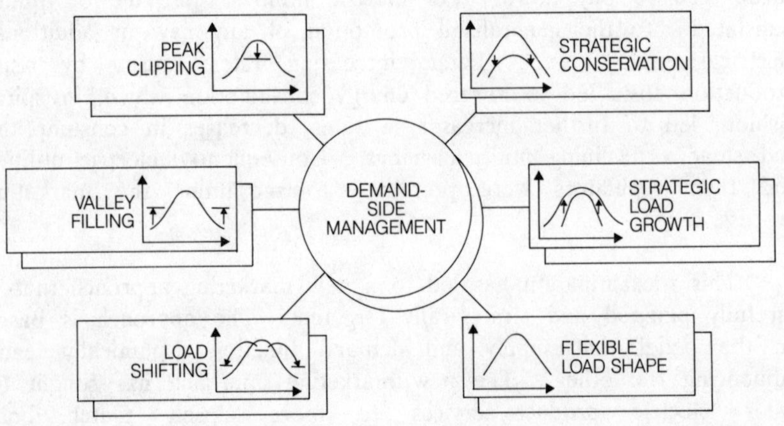

UTILITY DEMAND-SIDE MANAGEMENT FRAMEWORK

The basic framework for DSM planning is well documented and may be summarized as follows:

- **Specify objectives** -- Establish utility's desired goals.

- **Inventory alternatives** -- Identify the potential programs/strategies for reaching objectives.

- **Evaluate and select programs** -- Determine the evaluation process (decision methodology), evaluate alternatives, and select preferred alternative(s).

- **Implement selected program(s)** -- Conduct pilot-level experimental project or conduct full-scale program.

- **Monitor performance of program(s)** -- Measure success in achieving objectives, become aware of unanticipated "side effects" or contingencies.

This framework (shown graphically in Figure 2-2) describes DSM planning from the utility perspective. To date, however, utilities have not undertaken DSM planning in ways that adequately address the customer's perspective. Utilities have tended to view the customer solely in terms of his desire to minimize costs. This view is too simplistic. While an electric utility is concerned with sales price relative to product/service production cost, a customer is concerned with price relative to product/service value.

To a large extent, utility DSM program planning efforts have been directed toward products and services that affect utility load shape objectives. Relatively little research has been conducted on what utility customers perceive as needed and desired electric products and services. This is unfortunate. Customers make choices; they buy or do not buy specific products, subscribe or do not subscribe to specific services. And, having purchased or subscribed, they do or do not use products or services in specific ways. Changes in utility load shapes happen as the result of these customer actions. Consequently, customers' acceptance of, and responses to, utility programs have major impacts on whether the utility accomplishes its load shape objectives, and thus on the net benefits of the programs to the utility.

Figure 2-2. FRAMEWORK FOR DEMAND-SIDE MANAGEMENT PLANNING

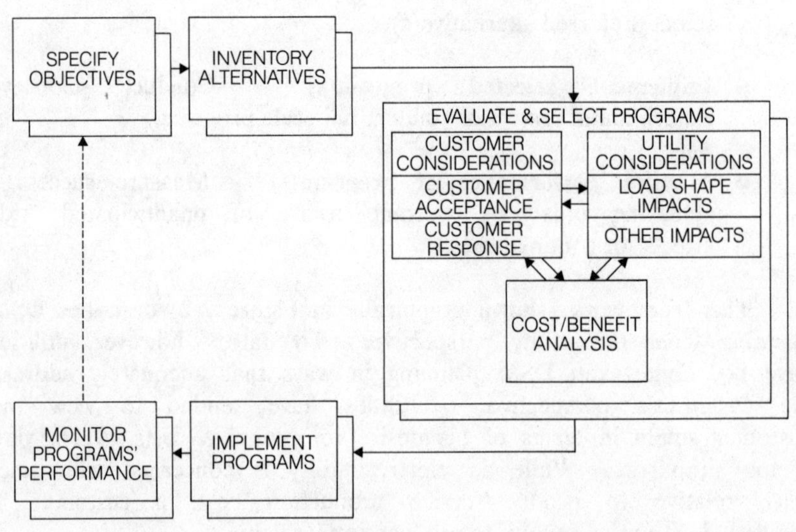

Utilities that seek to understand the customer's perspective and actively include it throughout their DSM planning increase the likelihood of creating successful DSM programs and reaching utility objectives. This chapter discusses the customer's perspective and how it can be applied to DSM planning.

THE CUSTOMER'S PERSPECTIVE: CUSTOMER DECISION-MAKING FRAMEWORK

In order to understand the utility customer's view of electricity purchase/use/consumption, it is helpful to examine an overall marketing model for consumer decision-making. The general decision process used by consumers in purchasing products is often outlined in consumer behavior theory in terms of four major stages:

- Problem recognition (unmet need)
- Alternative search and evaluation
- Actions (purchase[s]/participation)
- Post-action behavior (satisfaction?).

Changes in intrapersonal or external factors that result in discomfort or dissatisfaction cause a heightened attention/awareness and prompt the consumer to begin the decision-making process. The consumer becomes more receptive to (and may actively search for) internal and external stimuli/information in an attempt to understand the "problem." Problem recognition, the first phase in the decision process, occurs when the consumer believes there is a significant difference between the extent to which perceived or actual needs are fulfilled and the desired or expected fulfillment level.

Consumer Needs/Wants

Each person (consumer) experiences a set of basic and derived needs/wants (hereafter referred to as "needs"). When needs are not met to the desired or expected level (fulfilled), the individual is motivated to undertake actions that will help satisfy the needs. Basic needs include those experienced by most or all people, which exhibit a definable, limited range of variation among people and through time. Such needs are necessary to our survival and our ability to function as members of a social species. For example, whether from the 18th or 20th century, whether baby or adult, whether from China or the U.S., we need water, food, air, and some form of communication (i.e., language). An illustrative list of basic needs is given in column A of Table 2-1.

Derived needs are those that directly or indirectly originate from basic needs. Energy product/service needs are derived needs. An illustrative list of energy product/service needs appears in column B of Table 2-1. Fulfillment of a derived need functions to partially fulfill one or more basic needs. For example, heated water may be used to wash soil, bacteria, and perspiration from clothing (via a washing machine) and body (via a shower). Thus:

Table 2-1. BASIC AND DERIVED NEEDS

A Basic Needs	B Energy Product/Service Needs
Nutrition (food & water)	Warmed/cooled living space
Air	Hot water
Protection from environment (comfort)	Light
Protection from environment (security)	Accessibility (transportation and communication)
Sense of worth/individuality (identity, choice/control over one's life)	Work/production (process heat, machine drive, etc.)
Belongingness (includes communication ability)	
Health (freedom from severely debilitating mental or physical condition)	
Others' well-being	
Sexual satisfaction	

- Hot water helps clothes and body look and smell clean. Cleanliness (derived need) helps fulfill the need to belong (basic need), as most people prefer that their companions be reasonably groomed.

- Hot water also helps remove disease-causing bacteria from clothes and body. Decreasing the chances of disease helps maintain health (basic need). Heated water also helps prevent chill, reducing the body's thermoregulatory stress, and thus is also conducive to health and comfort.

Derived needs are not necessarily common to all people or constant through time. In other words, people seek to meet basic needs in different ways. For protection from severe summer heat (basic need) some people may prefer air conditioning, some, electric fans; 60 years ago some people depended on open-air porches, while others used big shade trees.

People also may seek to meet derived needs in different ways. Different people (or the same person in different years) may meet the derived need of heated water by purchasing a gas water heater, an electric water heater, or a solar-powered water heating system. The heated water example demonstrates how an energy product/service can be perceived by a customer as helping to fulfill more than one need (cleanliness, belongingness, health and comfort). The product/service is viewed as having multiple characteristics or attributes.

Although all people try to meet their needs, the relative importance (priority) of commonly shared needs differs among individuals. For example, let's say two people have similar financial resources and are trying to meet needs for security and comfort. The first, to whom security is more important, buys a fancy electric alarm system and a simple room air conditioner. The other, to whom comfort is more important, purchases an ordinary dead bolt for the front door and central air conditioning. Given finite resources (need-fulfilling power), individuals will commit more of those resources and commit them earlier to the needs and wants which, for them, take priority.

Finally, it should be remembered that because human beings are social beings, individuals do not necessarily make decisions in isolation. Interpersonal interactions are often a factor in decisions regarding basic or derived needs. Examples of multiple decision-makers are:

- Both spouses decide as a couple what brand and model of refrigerator to buy.
- Service person from a local appliance supply and repair store has an active say in what type of replacement air conditioners a small business owner chooses.
- Architect, building-owning partnership, and contractor decide together what space heating/cooling system goes into newly planned industrial building.

Consumer Influencing Factors/Descriptors

Many factors, both intrapersonal (individual-specific) and external, contribute to the shaping and definition of an individual's needs and perceptions of needs fulfillment. These factors can interact; i.e., they act as dependent variables that more or less affect each other. Illustrative lists of intrapersonal and external influencing factors are presented in Table 2-2.

Consumer Decision-Making Process

Let us use an example of a hypothetical Mr. Y to follow a consumer into and through the decision process.

Table 2-2. INTRAPERSONAL AND EXTERNAL INFLUENCES ON AN INDIVIDUAL'S PERCEPTION OF NEEDS

Intrapersonal Influences

Attitudes and knowledge based on past experience and education (information, emotions, learning, and memory)

Personality and self-image (e.g., does person see himself as a risk-taker?; does person perceive himself as inadequate or powerless?)

Informational processing (how individual screens, interprets, and uses stimuli, including time frame or horizon by which individual functions)

Habits

Physical characteristics: health (e.g., mental, physical), age, gender, race

Current and expected income and lifestyle (e.g., appliance use patterns)

Current stock, condition of goods (e.g., appliances), and level of services

External Influences

Cultural and sub-cultural, such as socioeconomic class or status, ethnic heritage, family or group traditions, geographic region (e.g., California or South Carolina, urban or small town/rural)

Interpersonal, such as family, friends, enemies, colleagues, acquaintances (some of these may function as a major party with customer in decision-making -- multiple decision-makers)

Commercial, such as advertising, salespeople, displays and packaging, accessibility of product or service

Hero figures/opinion makers (people or characters the consumer admires, envies, wants to be like), such as political figures, religious leaders, subject experts, superiors at work, sports stars, celebrities

Public, such as mass media, consumer rating or protection organizations, community-wide involvement

Natural environment (e.g., weather)

- **Problem recognition (unmet need)** -- It is a spring evening. As it gets later, the temperature drops (external influencing factor). Mr. Y's thoughts: 1) I'm uncomfortable. 2) Why am I uncomfortable? (Stage one: heightened awareness, open to information.) Conclusion: I am cold.* 3) I like being warm, not cold. (Desired level of need "comfortable body temperature" defined as warm. Being warm is one of Mr. Y's objectives -- desired level of need fulfillment.) 4) I have a problem that needs to be solved; how can I solve it? (Recognition of unmet need and active interest in searching for information on alternative ways to meet need.)

- **Alternative search and evaluation** -- Mr. Y looks for solution to problem, may search for and consider alternatives. He could go indoors and turn on the central heating (consume additional electrical energy). Or he could put on a sweater (conserve current energy). Or he could stop by a coffee shop for a cup of hot cocoa (consume energy from nonelectric source). He evaluates alternatives in relation to objective(s), which may include not only being warm, but also taking a walk with his wife.

- **Action** -- Mr. Y, being close to home, wanting to be warm, and wanting to walk with his wife, chooses to dash home, grab a sweater, and return outdoors (action).

- **Post-action behavior (satisfaction?)** -- Mr. Y notices the results of his action; he monitors the selected alternative. If he is still cold, he may return to the decision process and, this time, go home and turn on the heater. (Or, if his wife wants to continue the walk because she is enjoying the stars, she may ask him to stay and he may continue using the sweater -- an example of multiple decision-makers.)

---

*The authors recognize that people do not actually think this way. They simply feel cold and decide how to get warm. This example was chosen for purely illustrative purposes.

Several stages of Mr. Y's decision process may occur rapidly, with very little consciousness of thought. But how did Mr. Y evaluate the alternatives relative to his objectives, i.e., to his desired fulfillment of needs? Let us examine stage 2 of the consumer decision-making process, alternative search and evaluation, in more detail.

A person's satisfaction with life is a function of the fulfillment of the person's needs. And the fulfillment of needs is a function of the person's expectations regarding fulfillment of needs. In analyzing satisfaction, researchers have represented the relationship of value satisfaction to expectations and perceptions of the fulfillment of needs using a mathematical formula. Drawing from these researchers, a simplified formula expressing this relationship among energy service needs, perceived value, and expectations for any customer has been advanced as follows:

$$\text{Value} = \sum_{j=1}^{n} Wj \; (Aj - Dj)$$

Where:

Value is the customer's satisfaction with life

- n is the number of needs
- Wj is the weight or relative importance of fulfilling the customer's j^{th} need
- Aj is the extent to which the jth need is actually (or perceived to be) fulfilled
- Dj is the customer's desired or expected level of the jth need fulfillment

This equation incorporates the importance of expectations/ desires in the attainment of satisfaction; allows for the individual's ranking of different needs; and takes into account the impacts of current levels of needs fulfillment. The equation also shows that even if one need can't be met fully, a specific satisfaction level (Value) can still be achieved via fulfillment of other needs.

There are two strategies involved with this concept. One is a multiattribute product that makes up for not fully meeting one need by at least partially meeting another need; e.g., Mr. Y in his sweater may not be as warm as he wants to be -- his comfort is not satisfied. Yet the opportunity to take a walk with his spouse may fulfill his need for belongingness. Thus wearing a sweater promises Mr. Y enough satisfaction value that he chooses that option. The second strategy is a menu of products that complement each other in the extent to which they fulfill various needs.

In the model of the customer's energy decision process, the individual searches for ways to meet those needs that are relatively important (the Wj's are high; any $Wj >> 0$) and for which there is a wide negative discrepancy between the actual/perceived and expected/desired levels of need fulfillment (where $Aj - Dj << 0$).

For example, Mr. Y wants very much to be warm. And he is *very* cold -- which is a long way from warm. Applying the value equation, his condition might appear as follows. Let us say comfort is Mr. Y's need j = 2. To Mr. Y this need has relatively high priority (W2 is an 8 on a scale of 1 to 10). He defines the need for comfort as being met fully -- the desired state -- when he feels warm (warm = D2, is a 10 on a scale of 1 to 10). He perceives being cold as being something substantially less than warm (goose bumps and shivers = A2, is a 3 on a scale of 1 to 10). Then for j = 2, Wj (Aj - Dj) is 8 (3 - 10) = -56. Mr. Y is 56 value units less satisfied than if his comfort need were fulfilled. (A value unit is somewhat analogous to the "util" found in introductory economics courses.)

Once the consumer has evaluated alternatives and chosen one or more as preferred, action -- stage 3 in the decision-making process -- may follow. Yet many times action does not necessarily follow. Although the consumer intends to participate in a service program or purchase a product, barriers to implementation effectively delay, even block, action.

Some important barriers to actions are as follows:

- The consumer cannot get cooperation and lacks the ability to override other participants who hold opposing views in the decision-making process. Obviously, in cases of multiple decision-makers, this barrier can be a substantial one. For example, an industrial facilities operator wishes to implement duty cycling of air conditioning to reduce demand during the peak period on a time-of-use rate, but the facilities production manager strongly opposes the idea because of the potential negative impact such action might have on the workers.

- The consumer lacks adequate time or money to implement the desired action. For example, the local family grocery store owner may be too busy with food stocks, staff payrolls, customer complaints, sales forced by big chain competition, and government regulation-related paperwork to be able to take the time needed to obtain financing, locate and supervise a reputable contractor, and have an energy-efficient refrigeration system installed for produce, dairy, and meat counters.

- The consumer lacks access to the means of implementation. Access can be physical or informational. A service the consumer wants may not be offered in his or her community. An elderly or disabled person may lack the transportation needed to get to a hardware store for new furnace filters. A Spanish-speaking family may wish to weatherize their home, but because of language barriers. the obstacles to obtaining the necessary supplies and instructions are insurmountable.

If the consumer or external influences can prevent barriers from occurring or weaken barriers already present, then implementation of intended action is encouraged and facilitated. Consequently, the time period between decision intent and actual action can be lessened significantly.

Let's say that the barriers to implementation are overcome and the consumer takes action. Recall that Mr. Y dashed home and got a sweater to wear. As we noted earlier, the decision-making process does not end abruptly after implementation of the chosen alternative. The consumer monitors the consequences of his or her actions. In stage 4 of the decision-making process, the consumer, in essence, is asking, "Am I satisfied?"

Decisions entailing the altering of one's own or others' behavior involve stress. This is especially true when high levels of uncertainty or risk are perceived to be involved. Stress may take the form of unpleasant worry (extreme agitation, ulcers) or pleasant excitement. (We speak of everything from new ideas to weapons systems as being "exciting" because they invoke pleasant excitement responses.)

Post-action doubts or concerns are common for consumers. Such concerns can be strong when the implementation of a decision involves high costs, financial or otherwise; when several alternatives or a close second choice alternative have been forgone (opportunity costs); or when even the alternative chosen is viewed as a mixed blessing (that will bring significant future costs as well as benefits). The consumer wonders whether the information and feelings upon which the decision was based were adequate or accurate -- whether he or she did the right thing.

These concerns cause the consumer to be receptive to, or to actively seek, reinforcement that the decision made was a good one. The input of others becomes influential to the consumer, and as consequences of the action become apparent, the consumer's concerns are confirmed or dissipated.

If the consumer's doubts are alleviated (consequences of the decision are experienced as positive, Value increased), then the consumer exits the decision-making process and tends to form a habit. Such habit formation is particularly applicable when pleasant experiences are immediate or strong and clearly attributable to the decision, or to the positive feedback of others in response to the decision. Faced with the same or a similar problem in the future, the consumer goes through a truncated version of the decision-making process. For instance, the consumer may skip much of stage 2 (alternative search and evaluation) and move rapidly from intent into action. The action taken will be a repetition of the previous choice: same decision, same action, but with fewer post-action doubts.

For example, after considering his options, a customer for the first time joins in a direct load control program affecting his home air conditioner. Concerned about his decision, he closely monitors the thermostat, his comfort level, and his utility bills. The benefits of the program become obvious and soon he is very satisfied with the results. By summer's end, he feels that participating in the program was convenient and saved him money without significantly decreasing comfort. His neighbor, who also participated in the program, tells him that she liked it, too. Toward the next summer, he does not spend time and effort trying to decide what he wants to do; he happily and quickly signs up again for the direct load-control program. This season he does not peer owl-like at the thermostat every 20 minutes, as he did at the start of last season. He is confident of good results, so he does not monitor action results as closely as before.

If, however, the consumer's concerns are confirmed (consequences of the action are experienced as negative, satisfaction value decreased or increased less than expected), the consumer may continue to monitor the decision. Feelings of continuing or intense dissatisfaction, or changes in influencing factors, can cause the consumer to reenter the decision-making process.

If the above-mentioned customer had been dissatisfied with the effects of the direct control program, or if his children (important external influence) had complained that it was too hot, then he might well have reconsidered his choice and decided to leave the program at the earliest opportunity.

APPLYING THE CUSTOMER'S PERSPECTIVE TO THE UTILITY DSM PLANNING PROCESS

The utility plans its entire system and operation around its customers' expected demand for electricity. Any deviation in that demand from the forecast upon which those plans are based can create an imbalance in the costs and cost recovery of an electric utility. (This is an issue of note these days because plant siting difficulties and the uncertainty of recouping plant costs have become dilemmas common to many utilities.) In addition, the responses customers have to demand-side management programs, to the extent that they differ from expected responses, can also create such imbalances. When an imbalance occurs, changes in a forecast may result, leading to changes in capacity plans, revenue requirements, and DSM programs. These changes may, in turn, change the level and configuration of tariffs and the DSM alternatives offered.

DSM programs that are geared to maximize load control from the utility's standpoint, but which do not take into account customer needs, preferences, and behavior, are doomed to failure and create dissatisfied customers. These customers may eventually wind up as adversaries at hotly contested rate hearings -- or, in the case of large commercial or industrial users, as candidates for other, alternative sources of energy. The utility's ability to reach its objectives is placed at risk if it does not consider customer needs when implementing DSM programs.

Clearly, it is important that utilities consider the customer's perspective. A major step toward accomplishing this is to incorporate the customer's decision-making process into demand-side management planning. Understanding how this may be done is made easier by realizing that the utility's DSM planning framework and the utility customer's decision-making process are analogous, as shown in Table 2-3.

Table 2-3. ANALOGOUS PROCESSES

Utility DSM Process	Customer Decision Process
1. Specify objectives (where are we and where do we want to go?)	1. Problem recognition (where I am is not where I want to be: unmet need)
2. Inventory alternatives	2. Alternative search and evaluation (use of customer value equation as decision methodology)
3. Evaluate and select programs a. Determine evaluation process (decision methodology: impacts/criteria of interest, and decision rule) b. Evaluate alternatives c. Select preferred alternative(s) (including promotional strategy)	3. Action (purchase/ participation) 4. Post-action behavior/ satisfaction? (also includes appliance and service use patterns)
4. Implement selected program(s)	
5. Monitor performance of program(s)	

Recall that products and services are perceived by consumers as having multiple attributes. Each product/service, to a greater or lesser extent, affects the fulfillment level of many basic needs. Therefore, when a utility implements an energy product/service program, important issues are whether the utility intentionally affects customer needs, which needs, how these needs are affected, and to what extent. Ideally, the utility seeks to design and implement a program which significantly increases the value of the utility's total product/service mix to the customer. In other words, when taking the customer into account, the utility should seek to maximize "value" rather than simply minimize the customer's financial cost.

We can refer to the modeled customer value equation to understand ways the utility can increase value. In seeking to better match programs to customers, the utility can "reshape" the customers to fit the programs, or shape the programs to fit the customers. Following are some possible strategies.

Strategy 1 -- This strategy largely involves reshaping the customers, and is the typical method utilities have used to incorporate customers into DSM planning. It is a kind of "Now that we've chosen a program to meet our needs, how do we get customers to accept it?" approach.

We saw earlier that many intrapersonal and external factors shape a customer's needs and perceptions of needs fulfillment. The utility can act as an important external influencing factor and alter the characteristics of the customer's needs. This might be done via advertising or other promotional activities, newsletters, or other educational programs such as seminars and energy surveys (audits), and other forms of communication with customers (bill inserts, etc.). It has been argued that, for a number of electricity customers today, the incremental value of basic electricity service is not worth its price. For these customers, changes in either Wj relative to other Wj's, or decreases in Dj, could make the augmented incremental value of electricity greater than the selling price. Under such a condition, the customer would perceive the basic electricity service to be of net benefit.

For example, say a customer's need j1 is minimizing the price of energy. By showing the customer that energy is being produced in a cost-effective manner, and that the price charged the customer is only a small percentage higher than the utility's cost to produce energy, the utility, through education, may change the customer's expectation regarding what constitutes a low price for energy (Dj1), thereby increasing the customer's satisfaction.

Another example of this strategy, which took place in the late 1970s, involved the United States Department of Energy, which sponsored experiments using energy cost monitors, The monitors let energy consumers see directly the cost of using energy. When shown their electricity use on a daily cost basis, consumers concluded that the value of benefits received from appliance use was greater than the cost of the electricity. These experiment participants, pleased by what they now perceived as "cheap" electricity, increased appliance use and energy consumption. The Department of Energy, attempting to encourage energy conservation rather than increased consumption, changed the design of the systems to show what the monthly or yearly bill would be if the customers continued to use electricity at that day's level. When shown their energy use cost as a monthly or yearly bill, experiment participants perceived energy as being more expensive and subsequently cut back on appliance use.

The utility, as the influencing factor, can also function as an important data source. To achieve program success, the utility must effectively communicate the program's existence and benefits to its customers. For example, a utility can actively provide information to those customers who are searching for, and evaluating, possible alternative solutions to a problem. The utility does this by providing new customers with information packets containing brochures on the available utility programs in which they can participate. The utility that takes a role as information provider ensures that in the early stages of a decision, the customer (or potential customer) includes the utility's DSM programs among the alternatives to be considered.

The second and third strategies largely involve shaping programs to fit customers. These two strategies are based on incorporating the first two stages of the customer decision-making process into the first three stages of the utility DSM planning process. Stated differently, these strategies for including the customer's perspective in DSM planning are founded on two concepts: when specifying institutional objectives, utilities should consider the unmet needs of customers; and when identifying, evaluating, and selecting DSM programs, utilities should consider the decision process that the customer uses when evaluating such programs.

Strategy 2 -- The utility can plan a program to positively influence many of the affected customer needs. The following is an example of an energy service company that realized the potential of working with a local homeowners' association, and created an opportunity for increased revenue by designing and promoting a product/service package that addressed many of the association members' needs in a positive way.

The product offered: one brand of storm window. The product was presented not only in terms of its high quality materials and construction, but also as having multiple attributes that benefited many end uses. The product could: reduce utility costs by 15% to 17%; minimize dirt, dust, pollen, and insect invasion with full weatherstrip seal; maximize burglar deterrency; maximize comfort in all seasons; reduce outside noise by up to 50%; increase resale value of home; increase rental value of property; and reduce screen maintenance (new screens part of storm windows).

The services offered included: negotiating a bulk rate deal with the window manufacturer and passing on substantial savings to the association; getting the package qualified for the utility cash rebate program, with savings to go to the association; and selecting a reputable contractor and making sure that the windows were installed expertly before winter weather. These services clearly appealed to the association members' need for an economical program that was convenient and a low risk.

The result: The association, which had never previously given a thought to storm windows, is seriously considering what it believes to be a unique, beneficial offer. The association does not perceive the company as having any competitors. The company may win a sizable contract (272 condominium units, five or six windows each).

Strategy 3 -- The utility can plan a program to positively affect one or more of those needs that are most critical to the customer (and avoid negative impacts on other critical needs). For example, let's say a utility has received many complaints from residential customers. The customers' video cassette recorders did not record their favorite daytime television shows because voltage drops reset the VCR digital clocks. The utility might design DSM programs based on value of service that offer high reliability power to these and other customers, such as semiconductor manufacturers, who value a constant, dependable supply of energy during critical time periods. Such services could meet the customers' power reliability need and, simultaneously, meet the utility's need for increased revenues.

If a utility has a specific value or satisfaction formula (like the equation referred to previously) for each type of customer, the utility can evaluate how well each of several alternative demand-side management programs meets that customer's critical needs (customer's perception of the program's value), and can adjust the programs, or customer's price for participating in the programs, accordingly.

In another example, Narragansett Electric Company's standard service to industrial customers was unlimited, uninterruptible supply. The cost of such service exceeded its value to Rhode Island Forging Steel (RIFS), which wanted to locate a facility in an abandoned plant in Narragansett's service area. The utility considered both its needs and the critical needs of RIFS. Utility and customer negotiated off-peak business controllable and interruptible service at a new rate. The price was greater than the cost to the utility and was less than the value of the service to the customer. (The utility also benefited from the improved public relations climate fostered by satisfied customers, and achieved its desired impacts on load curve.)

Each energy product/service is designed by the utility to meet certain needs. For utilities, a key component of the demand-side planning decision process is the decision methodology (guidelines for evaluating alternative programs). In a simplified approach, the decision methodology can be specified by two items: the impacts or criteria of interest, and the prioritization, or weighting, used to determine the relative importance of the impacts (decision rule). Suppose the customer needs are defined as the impacts or criteria of interest. Further, suppose that the relative importance of needs to the customer is the decision rule. Then the customer's decision method becomes the utility's decision method for selecting demand-side management programs. The utility will choose the program that it predicts will increase the customer's net value or satisfaction the most. Expressing this in the format of the customer value equation:

$$\text{Value} = \sum_{j=1}^{n} Wj\ (Aj - Dj)$$

Where:

- Aj is the actual/perceived impact (on level of fulfillment of customer needs) from each program
- Dj is the desired impact (on level of fulfillment of customer needs) from each program
- Wj is the relative importance of the impacts (fulfillment of customer needs) for the customer

Obviously, the last two strategies for maximizing customer life satisfaction value can be achieved only imperfectly. No utility has the resources to examine all (including the most minor or peripheral) needs of its customers. Moreover, the utility must consider factors or objectives outside of the customers' concerns. Examples of such factors or objectives are:

- The utility's financial well-being, in both the short and long term

- Whether the utility can make a program accessible to intended program participants (e.g., ability to site a needed transmission line to the receiving neighborhood)

- The utility's strengths and weaknesses; i.e., the functions for which it is suited (e.g., the utility is not in the stereo supply business, but in order to meet customers' need for recreation it can supply energy to power stereos).

Customers' needs and wants and utilities' objectives and goals rarely overlap completely (Figure 2-3). Nonetheless, to the extent that the utility is willing and able to incorporate the customer's decision-making process into the utility's own DSM planning framework, the overlap of customer desires that are compatible with actively pursued utility objectives increases. Then customer acceptance increases and mutual benefits result.

Figure 2-3. REGION OF MUTUAL BENEFIT TO UTILITY AND CUSTOMER

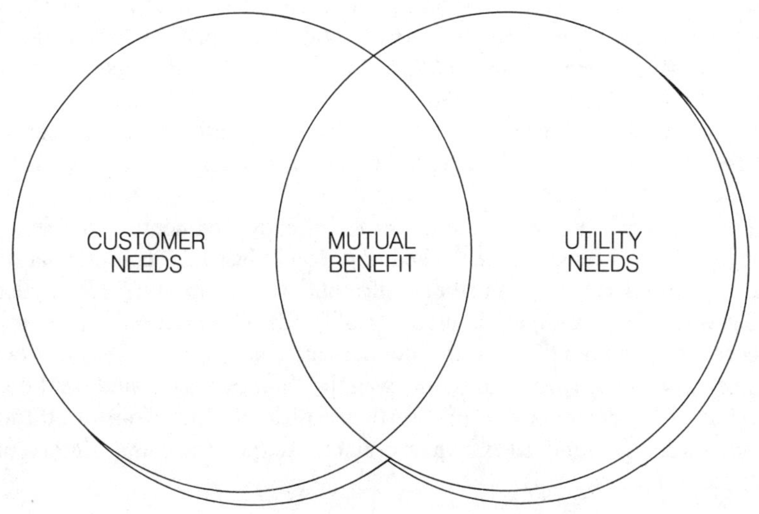

Of course, utilities have more than one customer, and it is not feasible for a utility to interview all of its customers about their needs. Fortunately, all of the aforementioned intra- and intercustomer variations are neither random nor hopelessly indecipherable. Individuals will have more or less similar and predictable derived needs, and priorities for basic and derived needs, to the extent that these individuals share common intrapersonal and external influencing factors. Certain factors or combinations of factors can be associated with certain needs. These factors can be used as descriptors; customers can be characterized by them. Individuals with similar characteristics will tend to have similar needs, and individual customers can thus be grouped according to descriptors. The utility can also segment its customer market by needs. It can offer each market segment programs that are both consistent with the utility's objectives and tailored to increase the segment members' perceived value of the service/product.

When establishing the decision methodology to be used in deciding which DSM programs to offer customers, market segmentation becomes an important issue. The utility should neither take every customer's decision method into account nor lump customers together into a single total customer or societal cost-benefit ratio. For most utilities, there really is no such thing as a market-wide average customer. Instead, in the DSM planning process, the utility should develop a utility decision method that is sensitive to the decision method of each segment's typical member.

To some extent, utilities have already segmented the market; they offer different rate schedules to industrial, commercial, and residential sectors. Further differentiation can be accomplished, however. For example, Southern California Edison used demographic descriptors to identify five residential customer segments called empty nesters, widows, young renters, wealthy homeowners, and middle-age singles. Segments were further divided by four "attitudes": disbelievers, economizers, marketplace supporters, and intervention supporters (Figure 2-4).

Figure 2-4. AN EXAMPLE OF RESIDENTIAL CUSTOMER SEGMENTATION

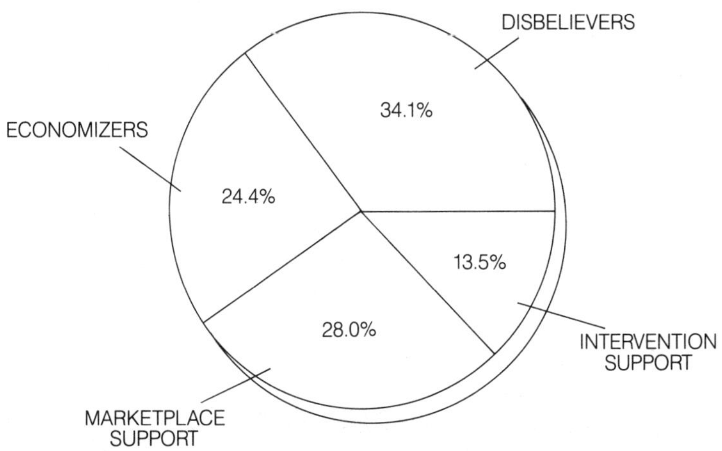

Disbelievers

- Do not reject cynicism
- Have little concern for future supply
- Disagree with need to do something to save energy
- Rate household conservation less important

Economizers

- Concerned with monetary savings from conservation
- Reject marketplace solutions to energy supply issues

Marketplace Supporters

- Accept marketplace solutions to energy supply issues
- Not concerned with social norms for conservation

Intervention Supporters

- Support legislative solutions to energy supply issues
- Oppose marketplace solutions for energy supply issues

Source: Southern California Edison Company

According to Whyte, "Demographic segmentation allows the planner to find people who have an opportunity to participate in the program; attitude segmentation allows planners to generate appeals by which these individuals are convinced to participate." The three strategies for increasing customer satisfaction value discussed above derive their effectiveness from sensitivity to the early stages of the customer decision-making process. Sensitivity to the later stages of the customer decision-making process -- action and post-action behaviors/satisfaction -- can also provide viable ways of incorporating the customer's perspective into the DSM planning process. When evaluating the implementation of DSM programs, utilities should consider how customers implement their decisions.

Strategy 4 -- The utility can design and implement its selected DSM programs to minimize barriers experienced by customers who are converting decision intent into decision action. For example, customers who want to upgrade worn-out hardware may only be able to choose from those models carried by area supply and repair companies. Consequently, if a utility wants to encourage conservation through use of energy-efficient appliances, it must first convince some of the area suppliers to carry energy-efficient brands and models before it can successfully encourage its customers to buy such appliances.

Other examples of utility customers facing barriers to decision action would be those living in a lower income neighborhood, who might be distrustful of a utility's program offering free home weatherization, or new immigrants who do not yet have a thorough grasp of English, and simply cannot understand the program. The utility can enlist the help of a local community group to communicate the nature of the program to residents and to actually carry out the weatherization.

Strategy 5 -- The utility can implement and monitor its DSM programs so as to encourage customer satisfaction with, and loyalty to, the decision to take part in such programs. As noted earlier, customers commonly experience doubts after a decision action. Utilities can help alleviate post-action stress in many ways. When planning DSM program implementation and monitoring, utilities should consider customers' post-action behaviors.

One way to minimize customers' doubts is to develop programs that minimize their perception of risk, lower or delay their costs, and increase, or hasten, their benefits. Zero or low interest loans and cash rebates for weatherization of buildings and energy-related hardware improvements are examples of conservation programs designed to maximize the customers' experience of benefits while minimizing their costs. Programs that include the leveling or averaging of a customer's bills also act to reduce the apparent cost and risk a customer experiences by reducing the "peak" month jolt to the customer's pocketbook and making energy use costs predictable.

Another way the utility can minimize post-action stress for the customer is by acting as a source of positive feedback. For example, after a customer becomes part of a program, the utility could follow up with one or more telephone calls, asking if the customer is satisfied with the program and addressing any customer concerns or questions. This direct, friendly contact helps the utility determine whether the customer feels the program meets his or her needs. At the same time, it increases the customer's perceived value of the service because the customer feels "included" and, therefore, valued by the utility (and the customer also feels smart for choosing to be the patron of such a considerate utility company!). This, in itself, may be enough to convince a wavering customer to stick with a program. In addition, the customer input may give the utility some good ideas on how to fine-tune its program.

In the above example, the utility determined customer satisfaction directly. The utility also can determine whether a voluntary program meets customers' needs through indirect means such as those described in the following examples.

- Do customers stay with the program or do they leave it? Customers with needs that continue to go unmet will be unsatisfied and inclined to leave the program. Satisfied customers will make their participation a "decision habit."

For example, a few years ago Pacific Gas and Electric Company noted that over 90% of one year's participants in its appliance cycling program for central air conditioners stayed with the program the next summer.

- Do customers who have the option stay with the utility or leave it? For example, the utility might measure dissatisfaction of large commercial and industrial customers with programs offered to this market segment by determining the number of segment members that "unplug" from the utility grid with self-generation, or plug into a neighboring utility's service.

CONCLUSION

Demand-side management planning has been viewed as giving only peripheral consideration to the customer's perspective. For utilities to take full advantage of the benefits DSM offers, they should incorporate the customer's perspective throughout the DSM planning process. In this chapter, we have described the customer's perspective and have discussed some of the possible strategies for including it in various stages of the DSM framework.

ACKNOWLEDGEMENTS

This chapter is reprinted, in part, with permission, from Proceedings of The Renaissance of Utility Marketing: Second Annual Seminar on Demand-Side Management (DSM), Electric Power Research Institute, Report RP2548-1.

The authors would like to thank B. J. Klosterman of UCI for her helpful comments on this article.

BIBLIOGRAPHY

1. Barrett, Karen. "For Love and Money." Ms., November 1985, pp. 43, 46, 48.

2. Battelle-Columbus Division and Synergic Resources Corporation. Demand-Side Management, Volume 1: Overview of Key Issues. Report prepared for Edison Electric Institute and Electric Power Research Institute, EA/EM-3597 Volume 1, August 1984.

3. Engel, J. F., D. T. Kollat and R. D. Blackwell. Consumer Behavior, 2nd edition. (New York: Holt, Rinehart and Winston, 1973.)

4. Engel, J.F. and Roger D. Blackwell. Consumer Behavior. (New York: Dryden Press, 1982.)

5. Flory, John E. (for Cyborex Laboratories, Inc.). Reflecting Customer Needs in Demand Management Planning. Electric Power Research Institute, Project TPS 82-652, June 17, 1983.

6. Gellings, Clark W. and John E. Flory. "Anticipating Changes in Customer Behavior, or Refocusing on the Customer." Electric Power Research Institute background paper for a presentation at the Energy Management and Communications Workshop of the American Public Power Association, October 18, 1983, Seattle, Washington.

7. Kotler, Philip. Marketing Management: Analysis, Planning and Control, 3rd edition. (Englewood Cliffs, New Jersey: Prentice-Hall, 1976.)

8. Lee, Fred N. "Minimum Cost Operation of Direct Load Control." (Dissertation, University of Kansas, Lawrence, 1983.)

9. Loudon, David L. and Albert J. Della Bitta. Consumer Behavior: Concepts and Applications. (New York: McGraw-Hill Book Company, 1984.)

10. Mazmanian, Daniel A. and Paul A. Sabatier. Effective Policy Implementation. (Lexington, Massachusetts: Lexington Books, 1981.)

11. Obeiter, Robert D. and Mary H. Smith. "DSM: A Case in Point." In workbook of Demand Side Management Conference: Market Acceptance, Load Shape Impacts and Planning Techniques, New Orleans, Louisiana, September 26-28, 1984.

12. Sanghvi, Arun P., Roger Levy and Joe B. Wharton. "Planning Demand Side Management Programs Based Upon Customers' Perceived Value." Preprint for IEEE-PES Winter Meeting, New York, February 3-5, 1985.

13. Wayne, Mary. "Demand Planning in the 80's." EPRI Journal, December 1984, pp. 6-15.

14. Whyte, M. D. "Development of Southern California Edison's Demand-Side Programs." In workbook of Demand Side Management Conference: Strategic Planning and Marketing, Dallas, Texas, June 21-24, 1984.

CHAPTER 3

Understanding Customer Preference and Behavior

Larry E. Lewis

INTRODUCTION

Electric utilities are facing many competitive pressures in the marketplace. Economic conditions are forcing utilities to undergo fundamental changes at a time when they face the double bind of uncertain sales and shifting economies of scale. The uncertainty of the timing and level of electricity demand has increased the risk of building new plants to meet growing demand. On the other hand, many electric systems have excess capacity and could profitably sell more power to the benefit of all customers. Regulators are reexamining traditional regulatory relationships, introducing the possibility of a more competitive price structure both within the service territory and between service territories. Traditional capital cost recovery methods and the well-defined boundaries of service may be permanently altered. Other fuel sources like natural gas offer stiff competition for electric markets. Customers who had always been considered captive now have relatively inexpensive technological options that allow them to self-generate or cogenerate. Customers leaving the system put an extra burden on the utility and its remaining customers by reducing revenue without an equal reduction in revenue requirements.

In order to survive in a more competitive marketplace, utility planners need a new way to think about marketing their products and services. Past marketing practices will not succeed in this new environment. Utilities can no longer follow a "selling" strategy based solely on providing a reliable product (kW). The customer has many options, and the evidence of the last ten years indicates that economic growth, particularly as the economy shifts further from industrial production to service provision, need not require ever increasing demands for electricity. What utility marketing planners need are marketing programs and services that meet company objectives, but also appeal to customers.

Demand-side management (DSM) -- planning and implementing utility programs designed to influence customer use of electricity and produce desired load shape changes -- has the potential to address the marketing needs and concerns of utilities. When DSM programs are successful, they benefit both utilities and customers. Customers have a number of needs or desires, which may include low energy costs, improved lifestyles, and a high quality of service from the utility. Utility objectives may include improved financial performance, increased return to shareholders, and improved customer relations. Studies have estimated that DSM programs could help some utilities reduce peak loads by 10 to 12% over the next 20 years, and increase off-peak loads correspondingly. The net impact of such programs could save customers approximately $100 billion through lower rates and deferred investment costs.[1] By holding down the cost of electricity in such a manner, DSM could strengthen the electric utility industry's competitiveness.

To realize their potential, DSM programs must achieve high levels of voluntary participation. Utilities have found that many customers choose not to participate despite financial incentives that appear to make participation economically attractive.[2] This suggests that electric customers are not driven solely by economic motives, contrary to popular belief. Customer decision-making is complex; it depends on customer needs and perceptions of the benefits obtainable from participation in DSM programs.

Marketing planners must develop an approach to understanding the customer, and target demand-side programs based on that understanding. They need to know how their customers evaluate energy service options and how different customer groups perceive the characteristics of each DSM program, product, and technology. With such a basis, utilities could:

- Design and market DSM programs that meet customer needs and preferences
- Estimate customer participation in these programs
- Target programs to customers whose participation would result in the greatest benefit for the utility and the customer.

Carrying out these three tasks could result in successful, cost-effective DSM programs that meet customer's needs as well as the utility's strategic objectives.

The basis for understanding the customer is preference and behavior analysis. Customer preference and behavior analysis uses various tools and techniques to produce a systematic understanding of how customers evaluate competing options and how they act on that evaluation. This understanding is used by marketing planners to design programs that will maximize marketing effectiveness within the context of the company's objectives.

Using customer preference and behavior analysis to maximize marketing effectiveness is not new. It is used successfully by a wide range of industries from dog food to television to automobiles to politics. Market research and marketing planning professionals have developed techniques for accurately identifying and measuring the frameworks and components of customer purchase decisions. These techniques are applicable to energy service purchases and can be powerful tools for improving the effectiveness of utility marketing efforts. This chapter will explore the value of customer preference and behavior analysis to utility marketing planners, describe the analytic method used to understand the customer, and offer an example of how such tools can improve marketing effectiveness.

THE VALUE OF CUSTOMER PREFERENCE AND BEHAVIOR ANALYSIS

In recent years, electric utilities in the United States have used demand-side management programs in response to critical issues facing the industry: growing competition from other energy providers, the high cost of new generating capacity, the need to hold down electricity rates, the need to achieve greater operating flexibility, the need to find new markets and products for electricity, and meeting strategic load and revenue growth objectives. The demand-side management planning process generally has started from the utility's overall company objectives, which are translated into load shape objectives. The utility evaluates the appropriate technology and non-technology options to produce the desired load shape change, and then designs a marketing program to encourage customer acceptance by appealing to the customer's energy service needs. In the conventional DSM planning process, the customer's energy needs and preferences are not considered until after the utility program has been implemented.[3]

If the utility marketing planner's mission is to satisfy the company's demand-side goals, then the planner must incorporate customer needs and preferences both at the beginning and during the planning process. The net result is a fully market-driven planning process that is influenced as much by the customer's decision process as by the utility's (Figure 3-1). The result of customer preference and behavior analysis is a more cost-effective selection and delivery of DSM programs, because the customer's decision process is linked to the utility's decision process.

In order to incorporate the customer's perspective into the planning process, the planner requires a complete understanding of the customer's decision process. The planner must be able to:

- Assess customer perceptions about programs and services offered by the utility and its competitors

Figure 3-1. LINKING THE CUSTOMER AND UTILITY DECISION PROCESSES

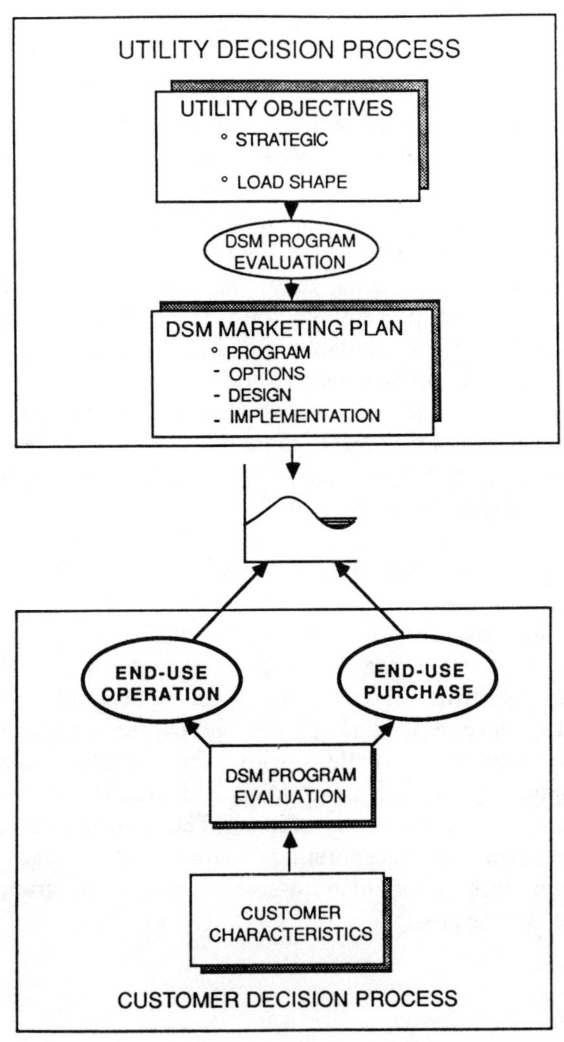

- Identify target markets and their defining characteristics

- Understand how customers make end-use purchase or program participation decision

- Isolate the reasons why customers do or do not participate in utility programs.

Fundamental to understanding the customer's decision process is customer preference and behavior research and analysis, which addresses the planner's most basic questions: how customers evaluate different energy service options; what types of media they prefer and are exposed to most often; how they prefer to learn about new products and services; whether they are motivated most by price or by their attitudes and perceptions toward the utility itself; how they process the information available to them and ultimately make a single choice of one option over another; the extent to which the utility can influence that process; whether customers can be grouped into segments within the overall market based on common purchase decision processes; and how marketing effectiveness can be improved by identifying such groups or segments.[4]

Customer preference and behavior analysis allows the utility marketing planner to develop a detailed DSM marketing plan in terms of program options, design, and implementation (Figure 3-2). The planner can start from the company's customer-driven demand-side management objectives, evaluate the most appropriate programs for influencing the desired load shape change by the customer, and then select the program attributes with the greatest likelihood of producing optimum program penetration and load shape impact for the least amount of marketing resources. The information also refines distribution choices by incorporating customer media and distribution channel preferences to broaden awareness of the program and make access to it as easy as possible.

Figure 3-2. DEVELOPING A DSM MARKETING PLAN

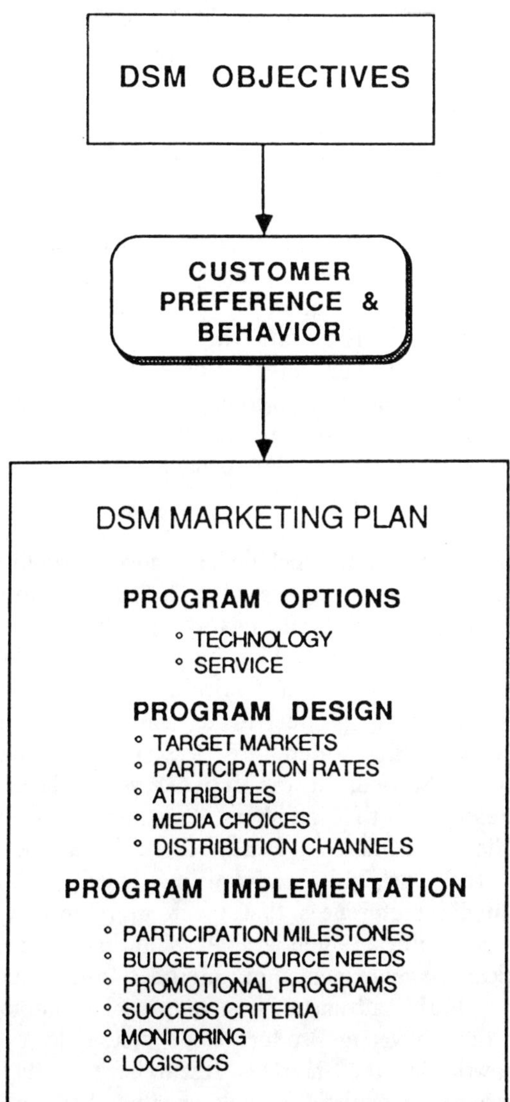

By understanding the market, marketing planners can define marketing opportunities, design demand-side programs that meet customers' energy needs more effectively than competing options, and maximize marketing efficiency by predicting how different groups of customers will react to different mixes of programs. This is the value of customer preference and behavior analysis.

ANALYZING CUSTOMER PREFERENCE AND BEHAVIOR

Electric utility planners, in particular demand forecasters, are familiar with the usefulness and accuracy of econometric and end-use models that estimate consumption by modeling appliance stock and consumption patterns. They are less familiar with the use of attitudinal and psychographic data. Some are skeptical that these difficult to quantify variables actually can be useful in utility planning. Yet the state of the art of market research is such that inclusion of customer preference and behavior data is commonplace in most industries.

The market research techniques now available to utility marketing planners permit analyses of complex behaviors that earlier had been impossible. Several factors in this development are important to note. First, the U.S. economy has evolved from a production-oriented economy to a service economy. Market research has changed its focus from product needs and wants to the needs and wants for services, which often are less tangible than those for production goods. Second, in the last 30 years there has been a significant increase in the numbers and types of new services available to the consumer. Third, there have been significant additions to the available advertising media, which makes understanding media preferences that much more important. Fourth, there has been a significant demographic shift in the structure of the "typical" American family; customer lifestyles and needs are more heterogeneous. Fifth, advances in computer technology and the introduction of the Universal Product Code system have produced an exponential growth in the market researcher's ability to process information, permitting statistical analyses that had previously been prohibitively expensive. Finally, developments in mathematical psychology techniques, which have benefited from the computer revolution, have produced more sophisticated measurement techniques for quantifying attitudes and preferences within smaller market segments.

Customer Purchase Decision Framework

Fundamentally, customers purchase products or services to satisfy some unmet need. Customers choose among options by evaluating the attributes of the options available to them. The evaluation process reflects the customers' internal balancing of beliefs, attitudes, and perceptions about the attributes of each option. The result of the evaluation is the formation of a preference for the one option among many that will best fulfill their unmet need. Provided the customer does not face constraints preventing him from acting on that preference, a purchase will be made.

The information processing literature in psychology suggests that there are five major stages of the purchase decision:

- **Problem recognition** -- identify unmet needs
- **Search for options** -- ways to satisfy needs
- **Evaluation of options** -- weighing of different attributes
- **Decision and implementation** -- purchase(s)/participation
- **Evaluation of decision consequences** -- satisfaction.

Figure 3-3 illustrates a conceptual framework of the customer purchase decision process. The utility customer's objective is to satisfy unmet energy service needs identified during the course of daily life. These needs can be as mundane as the need for hot water, air conditioning, cooking, and uninterrupted video recording or as specialized as maintaining "clean" power levels for the production of silicon chips. Generally, the energy service needs of commercial and industrial customers are more complex and varied than those of residential customers and often vary within and among business types and sizes.

Once the unmet energy service needs are identified, the customer begins to take inventory of his alternatives. These could be different fuel sources (gas versus electricity), different generation sources (cogeneration or self-generation), consumption during different times of the day (employing swing shifts or around-the-clock operations rather than the standard "nine to five"), or relocation to a different service territory to take advantage of rate differences.

Figure 3-3. CONCEPTUAL FRAMEWORK FOR THE CUSTOMER PURCHASE DECISION PROCESS

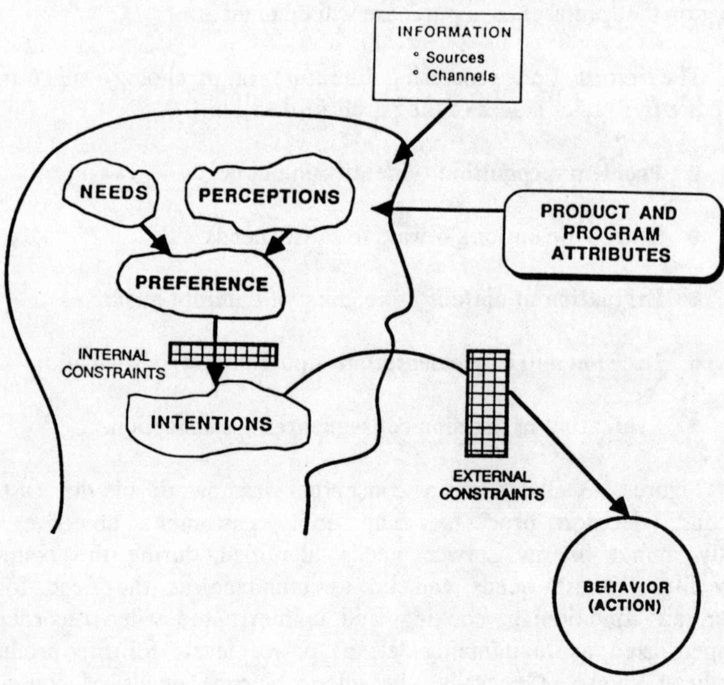

The customer then evaluates the different options and forms a preference for one over the others. Preferences are made up of beliefs, attitudes, and perceptions of the individual attributes of each option. A belief is a descriptive thought that a person holds about something. An attitude describes a person's favorable or unfavorable predisposition toward some object or idea. A perception is the process by which an individual selects, organizes, and interprets information to create a meaningful picture of the world. The important points for utility marketers are that preference elements can be influenced, and that the effect of those influences on preferences can be modeled and predicted.

The customer then moves from preference formation into the decision and implementation stage. The decision step can be characterized as an intention to act. The movement from intent to implementation, the purchase action, is affected by constraints, which can act as barriers to action.

After the purchase, satisfaction or dissatisfaction with the decision is fed back into the needs identification stage. If satisfied, the customer is likely to continue the same action. If his needs are unmet, the process will begin again until the need is met.

The power of customer preference and behavior information is its ability to identify and measure the factors that influence the customer purchase decision. Understanding these factors allows a marketing planner to predict purchase behavior given different program characteristics, which can result in marketing plans that can target specific programs to meet the needs of the individual customer. But in order to target effectively, the characteristics of the individual purchase decision must be attributed to specific groups or segments within the total market that have common preferences and behaviors.

In order to understand the markets for the utility's programs, the utility manager must acknowledge the relationship between customer characteristics and customer purchase behavior. The underlying assumption that binds this relationship is that customer characteristics drive customer purchase behaviors. Therefore, understanding of the market for electric utility programs is gained by analyzing customer purchase behavior in the context of customer characteristics.

In market research terms, customer purchase behavior is studied through choice modeling, while customer characteristics are studied through segmentation analysis. As individual modeling methods, choice modeling and segmentation analysis are useful research tools; however, integrated analyses offer unparalleled opportunities for understanding the dynamic relationship between purchase behaviors and customer characteristics, and the market as a whole (Figure 3-4).

Figure 3-4. ANALYSIS TECHNIQUES FOR UNDERSTANDING THE MARKET

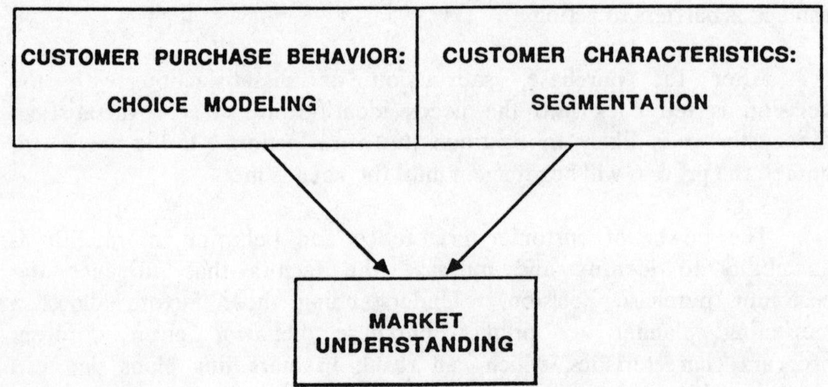

Choice Modeling

Traditional electricity market forecasting treated the customer as a "black box." Service attributes, socio-economic factors, market information, historical data, and constraints are inputs. Equipment purchase and consumption are the outputs. Choice modeling offers an improvement to this simplistic model by explicitly modeling the cognitive mechanisms in the "black box" that govern behavior. It emphasizes the step between option evaluation (preferences) and decision and implementation (behavior), which is critical to predicting the effectiveness of program design on customer participation.

Choice models are adapted from traditional economic models of customer choice, which presume that an individual's or homogeneous group's market behavior is motivated exclusively by a desire to maximize economic preferences -- the famous "rational economic man." This limited view of the customer motivated many demand-side program planners to emphasize price changes through rate or direct incentives to stimulate participation rates and increase the customer's preference for the program. However, the disappointing results of many of these programs show that the customer is not strictly an "economic man," and is driven to make choices because of other needs and preferences.[5] Choice models that implicitly incorporate attitudinal and behavioral data augment the econometric models by accepting that customer choice involves more than price considerations.

Choice modeling has four sequential analysis components: awareness/availability, perceptions, preferences, and choice. The customer must be aware of a program and have access to it, must perceive it as a realistic option, must form a preference for it over other options, and must choose to participate in the program. The analysis components are briefly described below.

Awareness/Availability Analysis measures the effectiveness of communication programs at establishing the customer's awareness of the options and the impact of different distribution techniques in making the program available.

Customer awareness is established primarily through advertising. Figure 3-5 illustrates the relationship between advertising spending and the level of awareness achieved for two hypothetical customer segments. The relationships form "awareness response curves" and are most often developed through the judgment of a marketing planner with experience in similar programs. Awareness can also be measured through direct field study. The analysis is used to assess the relative importance of media viewing habits, shopping habits, and information source utilization. The planner can then select the most appropriate media and retail outlets for advertising, promotion, and distribution, as well as the critical information sources for communication.

Figure 3-5. AWARENESS AND ITS RELATIONSHIP TO ADVERTISING SPENDING

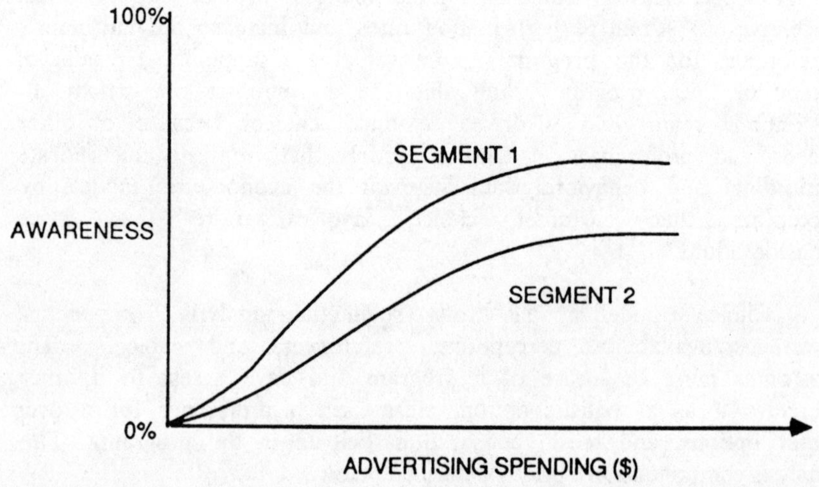

Although advertising can affect perceptions, preferences, and beliefs, the primary effect is in the attainment of awareness. Once a customer is aware that a demand-side program is available, he or she will decide whether to participate after rationally balancing the benefits that will be derived against the costs. Participation in many demand-side programs requires a financial or behavioral change commitment large enough to rule out "impulse" buying. The awareness and availability strategies of the utility planner, therefore, must be targeted to the customer most likely to be at a stage in the choice process where awareness of an attractive option will move him from intention to action.

Perceptual Analysis reflects the way individuals who are aware of programs perceive them relative to alternatives. Perceptual analysis provides a means, most often graphically, of portraying customer perceptions of product differences and the perceived attributes that characterize that product. Figure 3-6 illustrates one of the most common perceptual analysis techniques, multidimensional scaling, which plots customer perceptions of four different cost and energy efficiency product options. Without knowledge of the complete perceptual context into which the customer is integrating information, it may be impossible to predict the impact of an appeal to one attribute relative to another. For example, in direct load control programs where air conditioner owners allow the utility to cycle the unit during peak hours, it is important to understand how the customer perceives the value of his discomfort relative to the incentive offered by the utility in exchange for control of the appliance.

Perceptual analysis can also measure the customer's perception of the utility company itself versus other energy service providers. The dimensions might be cost, reliability, and concern for the customer as a valued "client" of the utility. By comparing customer perceptions toward the utility and its competitors, the marketing planner can focus marketing strategies on the perceived strengths and improve those attributes that the customer perceives as weaknesses relative to competitors.

Preference Analysis models how individuals in each market segment evaluate and rank alternative programs or products. Figure 3-7 shows an example of customers evaluating the same product, refrigerators, presented in several different attribute configurations. There are three principal analytic methods for measuring preferences for individual attributes: expectancy value, preference regression, and conjoint analysis. These methods identify the preferred attributes so that the most attractive mix can be offered. In many cases, utility marketing planners need to know how to select objective program dimensions. Of the three methods, conjoint analysis is preferred when objective program characteristics, such as incentive amounts and appliance efficiency levels that qualify for a rebate, are the focus of the design problem.

Figure 3-6. PERCEPTUAL ANALYSIS GRAPHICALLY DEPICTS A CUSTOMER'S PERCEPTIONS OF DIFFERENT PRODUCT ATTRIBUTES

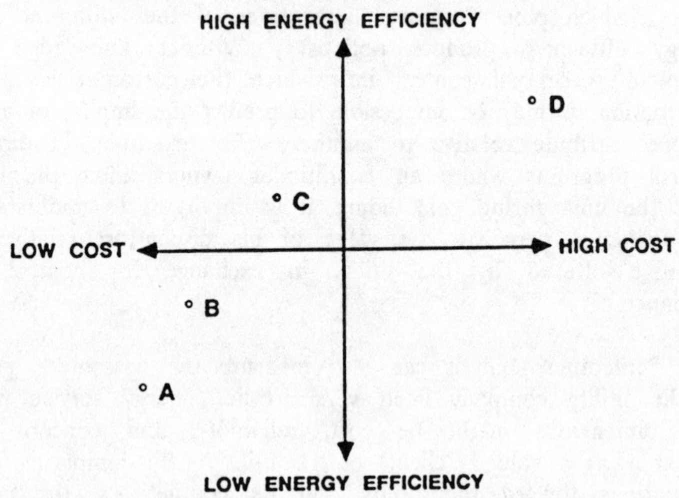

PRODUCT DESCRIPTIONS

A: Energy Inefficient A/C System Brand 1
B: Energy Inefficient A/C System Brand 2
C: Energy Efficient A/C System
D: Solar A/C System

Figure 3-7 illustrates a conjoint analysis. The rating of each configuration package can be disaggregated into ratings of the individual product attributes such that a set of values for each attribute level can be developed. This value set can be made for each market segment of interest. This process is fundamental to identifying the attributes most likely to help the customer form the intention to purchase a product or participate in a program.

Figure 3-7. PREFERENCE ANALYSIS IDENTIFIES WHICH ATTRIBUTES THE CUSTOMER PREFERS OVER OTHERS

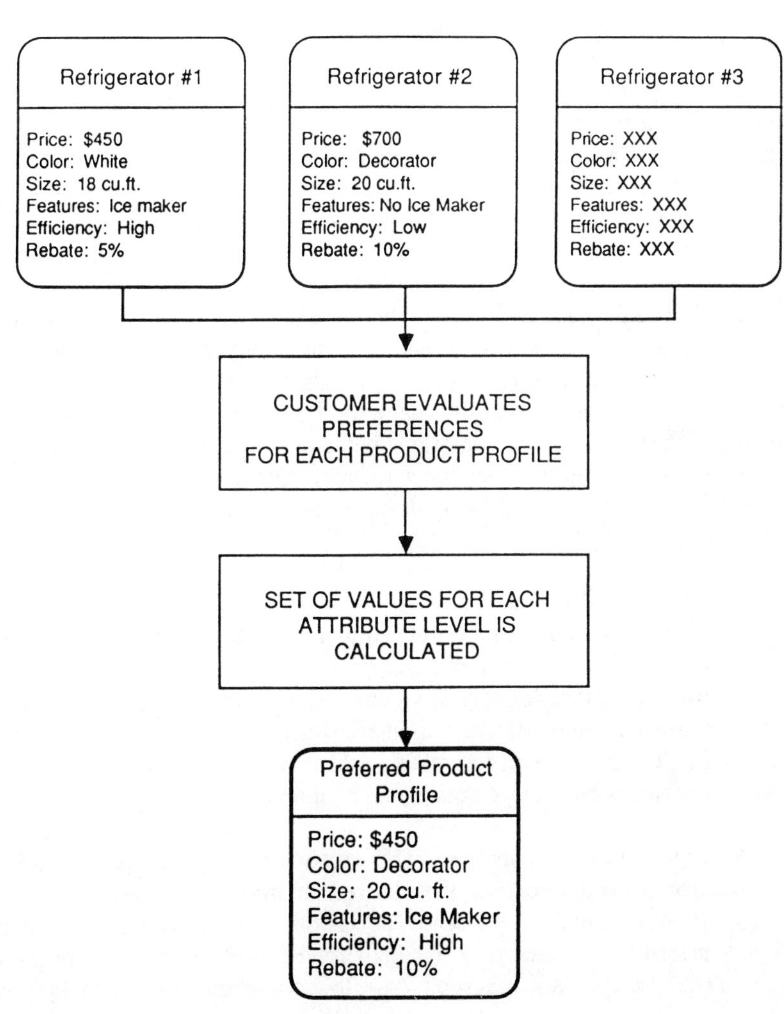

Discrete Choice Analysis translates preferences into choice. It measures the critical link between program design (preferences for specific attributes) and the market share impact within each segment (purchase behavior). The analysis presumes that a customer will choose the product with the highest perceived value based on his preferences for different attributes of a product or program. With the awareness/availability, perceptual, preference, and choice analyses, the marketing planner can configure a variety of product or program mixes and estimate the likely purchase behavior for any given customer or customer segment with like preferences. Figure 3-8 illustrates that a discrete choice model can predict behavior in the form of market share percentages for a series of hypothetical market segments.

A thorough choice analysis should also include an investigation of the barriers to choice. This analysis refines the market share estimates from the discrete choice analysis by accounting for external forces that could influence the likelihood that participation estimates will be reflected in actual market behavior. Also, the barriers can accelerate or decelerate the time it takes for the segment to reach the predicted level of participation. Some important barriers to action are:

- **Multiple decision-maker** -- A customer wishing to participate in a program cannot get cooperation from others who also have influence on purchase decisions and lacks the ability to override other opposing views. This barrier is most prevalent among commercial and industrial customers and has been seen in demand-side programs that have been accepted by building operators but delayed or vetoed by management.

- **Inadequate resources** -- The customer lacks adequate time or money to implement the desired action. There is insufficient capital available, or other decisions and needs have a higher priority. There may be paperwork and approvals necessary that seem so onerous as to outweigh the benefits of purchasing the service or appliance.

Figure 3-8. DISCRETE CHOICE ANALYSIS PREDICTS THE LEVEL OF PARTICIPATION BY SEGMENT FOR A PROGRAM GIVEN ITS SPECIFIC ATTRIBUTES

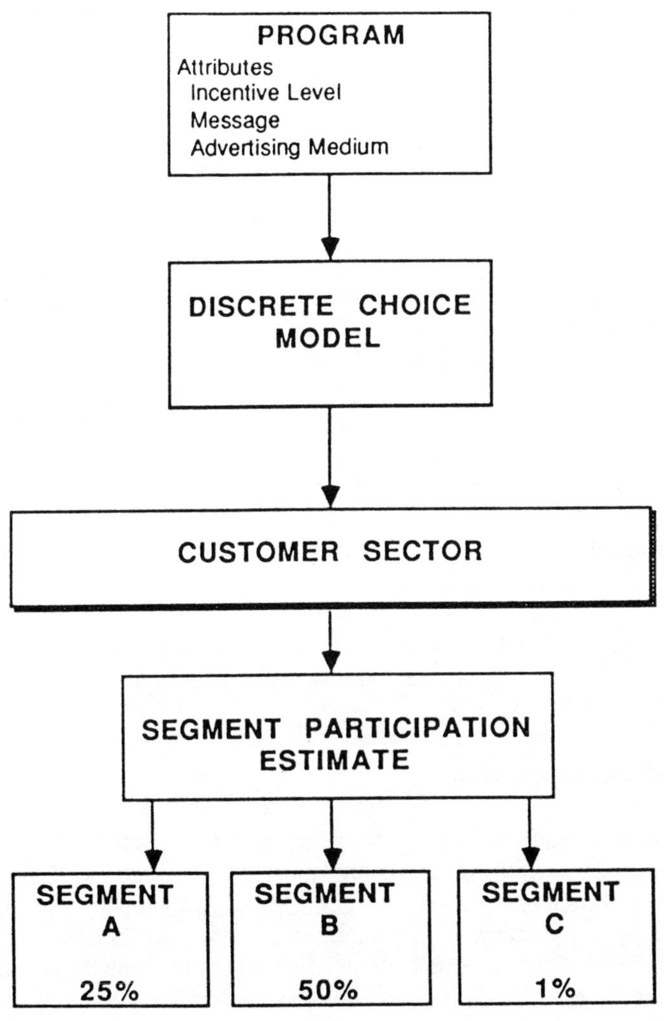

- **Awareness of availability and access** -- The customer lacks access to the means of implementation. Access can be physical or informational. For example, a program a customer wants to participate in is not offered in his community. One eastern utility offering a rebate for an appliance with a certain efficiency level found no takers. It was only later that they discovered that no vendor in the area carried the qualifying appliance.

Utility demand-side planners have recognized these barriers and have employed customer education, financial incentives and low-interest loans, trade ally cooperation, direct contact, audit services, and innovative rates to overcome them. Figure 3-9 shows that differing levels of advertising intensity can affect availability and access, and thus the length of time needed to reach ultimate program penetration. What is important is the recognition that customers make purchase decisions within the context of these barriers and that marketing measures can be taken to overcome them. Recognition of these facts is critical to program budgeting decisions and identification of the point of diminishing returns for continued marketing efforts.

Being able to predict the success of specific demand-side programs based on preferences for particular program characteristics is valuable only if groups of customers with like purchase behavior can be identified. The method of identifying those groups is called market segmentation, which allows target marketing by specific characteristics and preferences.

Market Segmentation

Market segmentation is a synonym for classification. Market segmentation was introduced to the marketing field in a classic 1956 article by Wendall Smith in the Journal of Marketing.[7] Implicit in the concept of market segmentation, and its practice, is the conviction that market composition need not, and indeed should not, be happenstance. Market segmentation's underlying premises are that variations among customers are systematic, and that correct classification provides guidance to cost-effective marketing that matches customer needs better than generalized approaches, thus increasing sales or market share. In its broadest application, it is an overall strategy for allocating market resources.

Figure 3-9. USING ADVERTISING INTENSITY TO OVERCOME PROGRAM PARTICIPATION BARRIERS

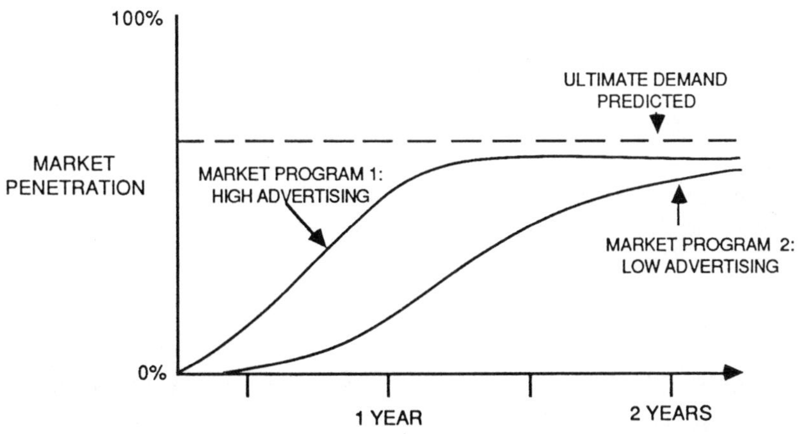

Source: Booz-Allen and Hamilton[6]

The reason for the growing utility interest in segmentation techniques is the historical evidence indicating that traditional segmentation of utility customers -- residential, commercial, and industrial -- is insufficient to optimize marketing resources. Even when utilities have broken down the commercial sector into segments defined by demand level or building square footage, the results for demand-side programs have been unsatisfactory. By understanding market segment characteristics, utilities can increase program acceptance. For example, planners may want to predict participation in an audit program based upon a segment's attitudes and buying behavior toward other products.

Accurate segmentation also can help the marketing planner to find that level of market disaggregation that will produce the most cost-effective marketing effort. For example, an all-encompassing marketing program that involves sending one type of program announcement to all residential customers may be less expensive than other media, but will probably be less effective. Conversely, customizing a program announcement for each individual residential customer will likely be more effective but could also be prohibitively expensive. Figure 3-10 illustrates the continuum of cost-effective marketing given the level of market segmentation.

Segmentation allows the utility marketing planner to increase sales or market share cost-effectively by virtue of one or more of the following tactics:

- Designing products or programs to meet the specific needs or desires of high volume or high potential customers

- Designing prices or incentives to attract customers at a maximum level of return to the utility

- Positioning products or services with regard to benefit-cost perceptions of the customer

- Advertising or promoting products or services through particular media or particular channels of distribution.

Segmentation is a commonplace activity. We all engage in our own classifications with a frequency that often escapes our conscious attention. We divide our friends between single and married, with children and without (demographic, socio-economic), and think of the possible activities we might engage in socially taking those classifications into account. When we buy gifts for friends we segment by hobbies and interests (attitudinal).

Figure 3-10. THE LEVEL OF MARKET SEGMENTATION AFFECTS THE COST-EFFECTIVENESS OF MARKETING EFFORTS

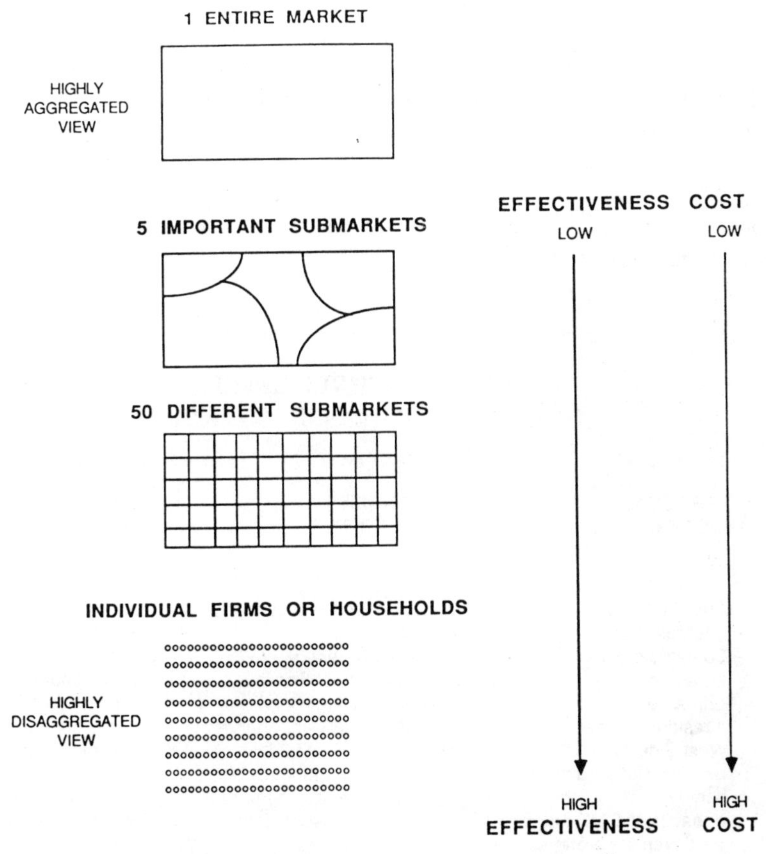

Source: Chakravarti, Hendrix, and Wilkie[8]

Many industries use segmentation. An obvious example is the automobile industry. General Motors was originally divided into separate divisions -- Cadillac, Pontiac, Buick, Chevrolet, Oldsmobile -- specifically to match segments of the American buying public based on income. Today, the segmentation has become very sophisticated. There are hundreds of colors, options, and body styles available to meet the multitude of auto buying segments. More importantly, the advertising often strives to meet lifestyle expectations rather than realities. They know that many consumers use their car as a statement not only of where they are socially, but where they want to be.

There are literally hundreds of ways to classify customers. However, these reduce to the three basic levels of classification depicted in Table 3-1:

Table 3-1. THE THREE LEVELS OF CUSTOMER CLASSIFICATION

LEVEL ONE	LEVEL TWO	LEVEL THREE
"PERSONAL CHARACTERISTICS"	"BENEFITS SOUGHT"	"BEHAVIORAL MEASURES"
Mary Winthrop is: - a female - 36 years old - an attorney - unmarried - a resident of the West Side - an owner of an older home - a reader of TIME magazine - a convenience-oriented shopper - an "affluent"	Mary Winthrop places high importance on: - Level of Service - Continuity of Service - Availability of Service She places low importance on: - level of her monthly bill - moderate rate increases - energy conservation	Mary Winthrop: - is a heavy user of electricity - does not own a heat pump - has not had an energy audit - has purchased two insulation items - is favorably inclined toward her utility

Source: Chakravarti, Hendrix, and Wilkie[8]

- Personal characteristics
- Benefits sought
- Behavioral measures.

Table 3-2 shows some representative measures for residential customers within each level. Table 3-3 shows an example for commercial or industrial customers. Utility marketing strategies can influence the benefits sought by designing programs that reflect customer needs, and have a greater influence over behavioral measures by helping the customer to form preferences and intentions to purchase products or participate in programs.

The following example illustrates how utilities can use market segmentation to plan marketing strategies. A recent utility survey attempted to identify factors that could be used in residential market segmentation by examining the needs and benefits associated with the purchase and use of electric energy and appliances.[9] The objectives of the market analysis were: identification of the needs and benefits that determine interest and participation in utility load management programs; differentiation of groups, or segments, on the basis of their concern with particular needs and benefits; and description of other characteristics of segments that could assist in marketing load management programs to them. A key question was whether the appropriate factors would be specific to individual appliances or end uses, or associated with energy and appliance use in general. This is crucial because if customer needs and benefits are specific to particular appliances, then marketing strategies based on segment buying behaviors would have to be designed specifically to each appliance. If, however, the needs and benefits are general across end uses or utility programs, marketing strategies could encompass a wider variety of programs.

Table 3-2. REPRESENTATIVE MEASURES FOR RESIDENTIAL ENERGY SERVICE PURCHASERS AT EACH CLASSIFICATION LEVEL

PERSONAL CHARACTERISTICS	BENEFITS SOUGHT	BEHAVIORAL MEASURES
Demographics		Energy Use
- Household Age	- low price	- ownership of appliances
- Age	- efficiency	- type of heating system
- Marital Status	- compactness	- type of cooling system
- Education Level	- heat evenly	- demand characteristics
- Income	- dependability	
- Housing Status	- conserves energy	
- own vs. rent		Energy Conservation
- location		- program participant
- value		- specific actions taken
- neighborhood		
- age		
Media Habits		Predispositions to Future Behavior
- newspaper readership		- intention to purchase
- radio station listenership		- intention to participate
- magazine readership		- attitudes toward:
- television viewing		- usage options
		- conservation
"Psychographics"		General Attitudes
- customized inventories		- toward utility
- syndicated inventories e.g., VALS		- toward energy use
		- toward conservation
		Others

Source: Chakravarti, Hendrix, and Wilkie[8]

Table 3-3. REPRESENTATIVE MEASURES FOR COMMERCIAL/INDUSTRIAL ENERGY SERVICE PURCHASERS AT EACH CLASSIFICATION LEVEL

COMPANY CHARACTERISTICS	BENEFITS SOUGHT	BEHAVIORAL MEASURES

Business Category
- commercial vs. industrial
- SIC code
  - retail
  - financial
  - manufacturing
  - high technology
	- efficiency
	- cost-effectiveness
	- reliability
	- dependability
	- cleanliness
	- safety
		Energy Use
		- equipment in use
		- type of heating system
		- type of cooling system
		- demand characteristics
		- others

Size of Firm - number of employees - volume of sales/production - other measures		Energy Conservation - program participation - specific actions taken

Location
- urban/rural
- proximity to transmission lines
- free-standing vs. attached
		Predisposition to Future Behavior
		- intentions to purchase
		- intentions to participate
		- attitudes toward:
		  - usage options
		  - conservation
		- awareness/understanding
		  - options
		  - programs

Performance Trends - favorable/unfavorable		
Ownership - chain vs. independent - own vs. lease		General Attitudes - toward the utility - toward energy supplies - toward conservation
"Psychographics" - innovativeness - risk oriented		

Source: Chakravarti, Hendrix, and Wilkie[8]

The factors identified were largely independent of specific appliances and end uses. Seven underlying residential needs/benefits factors emerged:

- Search minimization
- Electricity conservation/budgetary concerns
- Personal control
- Safety concerns
- High-tech enthusiasm
- Task lighting
- Appearance.

After the analysis isolated seven needs/benefits factors, customers were grouped based on the similarities and dissimilarities of their concern with each factor. On the basis of this examination, the study distinguished six segments, or clusters of people who have similar energy needs to others in the cluster, but who differ in other ways from people in other clusters. Three of the six segments are used in Table 3-4 to illustrate the clustering: economizers, value-seekers, and care-free customers.

Finally, the study compared the members of the six clusters with respect to other characteristics that can help marketing planners to address three important marketing questions: What is the best market toward which to target load management programs (Table 3-5)? What message will have the greatest appeal? And what is the most effective media channel to use (Table 3-6)?

Table 3-4. CLUSTERING CUSTOMERS BASED ON THEIR NEEDS/BELIEFS

Segments	General Needs/Benefits Factors						
	Search Minimizaton	Electricity Conservation/ Budgetary Concerns	Personal Control	Safety concerns	High-Tech Enthusiasm	Task Lighting	Appearance
Economizers	---	+	--- ---	O	O	+	O
Value-Seekers	---	O	O	O	O	---	---
Care-free Customers	++	--- ---	+	---	---	O	O

++ Very High + High O Neutral --- Low --- --- Very Low

Table 3-5. ATTITUDES TOWARD LOAD MANAGEMENT

Segments	Probability of Participation	Attitudes
Economizers	●	Open to load management devices for cautious economizing.
Value-Seekers	◍	Amenable to load management devices if incentives are offered.
Care-free Customers	O	Little interest in load management but will participate in utility if easy to do so.

High Probability ● Medium Probability ◍ Low Probability O

Table 3-6. BEST MARKETING MEDIUM

	Marketing Medium				
Segments	Mass Media	Consumer Publications	Bill Inserts	Direct Contact	Trade Ally
Economizer	very effective	very effective	effective	ineffective	ineffective
Value-Seeker	ineffective	effective	very effective	ineffective	effective
Care-free Customer	ineffective	ineffective	effective	very effective	ineffective

very effective (●) effective (gray ●) ineffective (○)

AN APPLICATION

By integrating choice modeling and market segmentation, marketing planners can identify and target markets for particular demand-side programs. The following example, using heat pumps, illustrates the utility application of the customer preference and behavior analysis for obtaining program participation estimates. (For a more comprehensive review of utility efforts to market residential heat pumps, see Chapter 13.)

In this heat pump example, the marketing planner would first make participation estimates using a set of base program attributes. Table 3-7 shows the base program attributes and the estimated participation percentages for each of the three segments. The next step would be to look at the needs/benefits factors provided in Table 3-4, focusing on the factor or factors that most distinguish the segments. "Electricity conservation/budgetary concerns" is the factor with the most diverse cluster of needs/benefits. The utility planner would then change the program attributes that affect this factor most directly. For example, a heat pump with low first cost could be promoted, installation could be included, the warranty period could be extended, and a discount could be provided. The planner then would re-estimate participation for each segment given the new program configuration, as shown in the modified program in Table 3-7. Figure 3-11 shows the percentage changes in participation estimates between the two programs. By looking at the changes in the participation, the market planner will be able to evaluate the program changes. Figure 3-11 shows that the segment that would be most responsive to the program changes is the economizers. This is the segment the utility should target first.

Table 3-7. HEAT PUMP PROGRAM PARTICIPATION ESTIMATES

Program Attributes	Segments	Participation (%)
High Equipment Cost Installation Cost= $300 5 Yr. Warranty No Discounts	Economizers	20
	Value-Seekers	30
	Care-free Customers	10

A Modified Program

Program Attributes	Segments	Participation (%)
Low Equipment Cost Installation Cost Included 20 Yr. Warranty $20/month Discount for 3 yrs.	Economizers	35
	Value-Seekers	32
	Care-free Customers	10

Figure 3-11. PERCENTAGE CHANGES IN HEAT PUMP PROGRAM PARTICIPATION

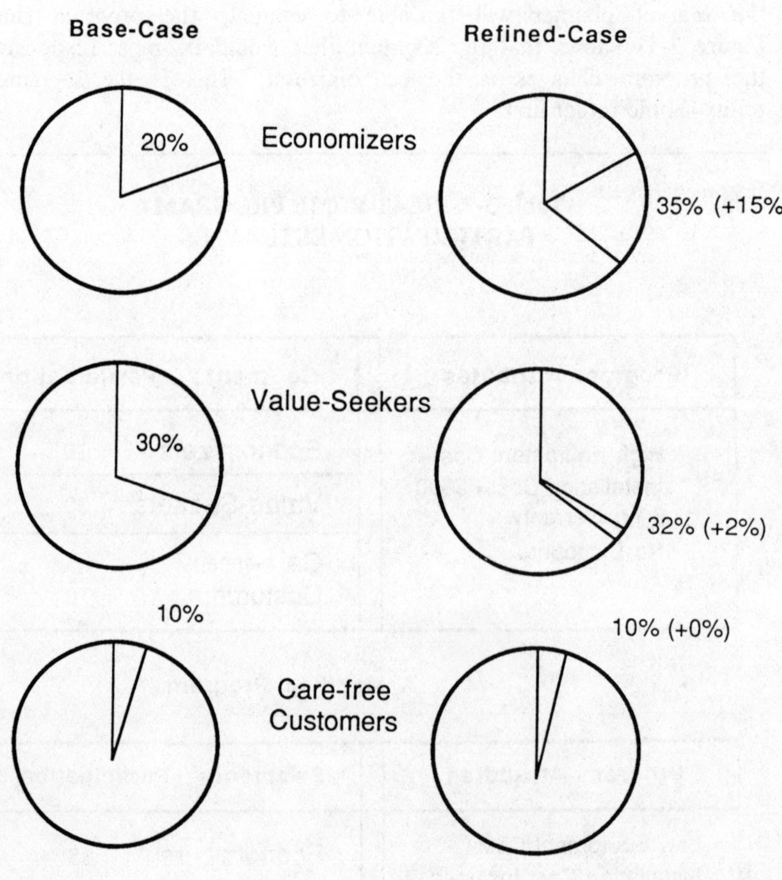

This application example illustrates how customer preference and behavior analysis can refine program design and participation estimates. More information and analysis would be necessary to complete a marketing strategy for a specific program for a specific utility. However, this example gives the flavor of how customer preference and behavior tools can help to target markets and refine participation estimates.

CONCLUSION

Utility industry efforts to use customer preference and behavior analysis will be evolving over the next few years. Yet there are many useful insights and benefits utilities can gain now if the tools and techniques are understood and applied. Customer preference and behavior tools are available now to utility marketing planners,[4] and can be used to respond to competitive pressures through the design of demand-side programs that meet customer needs and preferences. With these tools and techniques, utilities will be able to improve marketing effectiveness while satisfying both their customers' energy service needs and their company's strategic objectives.

REFERENCES

[1]Gellings, C. W. and T. Keelin, Impact of Demand-Side Management on Future Customer Electric Demand, Electric Power Research Institute, RP 2381-4, November 1986.

[2]Xenergy, Inc., Customer Acceptance of Demand-Side Management Programs: Commercial Sector, Interim Report, EPRI Project RP 2548, March 1, 1986, and Synergic Resources Corporation, Guidebook on Residential Customer Acceptance: Designing Programs That Attract Customers, EPRI Project RP 2548, June 28, 1986.

[3]Wayne, M., "Understanding the Consumer," EPRI Journal, October 1986, pp. 1-3.

[4]Electric Power Research Institute, Customer Preference and Behavior: Project Overview, RP 2671, November 1986, pp. 1-5.

[5]Lewis, L. E., "The Case for Behavioral Energy Modeling," presented at the Families and Energy Conference, Lansing, Michigan, October 1983.

[6]Booz-Allen and Hamilton, Inc., "Customer Preference and Behavior Project: Task II -- Modeling and Data Collection Methods," prepared for the Electric Power Research Institute, RP 2671, October 10, 1986.

[7]Smith, Wendall R., "Product Differentiation and Market Segmentation as Alternative Marketing Strategies," Journal of Marketing, July 1956, pp. 3-8.

[8]Chakravarti, D., P. E. Hendrix, W. L. Wilkie, Market Segmentation Research, Monograph 3, Volume 1, Market Research Monograph Series, Electric Power Research Institute, January 1987.

[9]Feldman, S., C. T. Finkbeiner, J. Berrigan, and L. E. Lewis, "Residential Segmentation for Marketing Utility Programs: General or Specific," presented to the ACEEE Summer Study Conference, University of California, Santa Cruz, August 1986.

CHAPTER 4

Anticipating Changes in Customer Behavior, or Refocusing on the Customer

Clark W. Gellings and John E. Flory

INTRODUCTION

Electricity users are unhappy. Many believe they are not getting their money's worth from their electric utility, and yet it is these unhappy customers who must ultimately provide the money that runs utilities.

Attitude surveys have shown that, until just recently, the percentage of customers who believe their electricity rates are too high was on the rise.[1] The survey results shown in Figure 4-1 reveal that the percentage of customers believing their electricity rates are "very high" or "somewhat high" has increased from 30% in 1971 to 50% in 1978 to almost 70% in 1982. Fortunately, rate increases have abated and people are not as upset as they once were. However, competition for large use customers is increasing, and the loss of such customers raises the specter of an unwanted shift of much of the revenue burden to smaller customers, as happened in the telecommunications industry. Utilities do not want to rekindle the belief that rates are too high.

What can utility managers do to heighten the customers' perception that they are getting their money's worth from the electric utility? One often neglected solution is to refocus on understanding the customer's behavior, needs, and preferences. Electricity service can be adjusted according to actual customer preferences once these preferences are understood, and the customer will feel increasing satisfaction with the service.

Figure 4-1. TRENDS IN CUSTOMER ATTITUDES TOWARD RATES

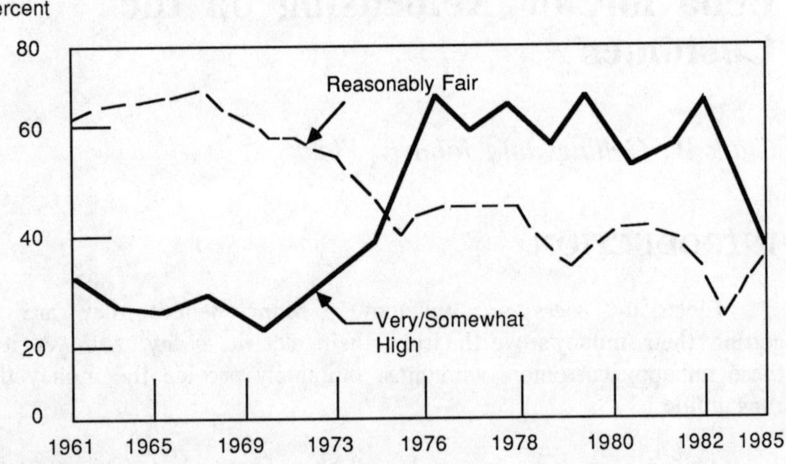

Source: Cambridge Reports Inc.[1]

Refocusing on the customer will help utility managers in three areas: forecasting, energy service planning, and customer relations. Improving work in these three areas helps give customers their money's worth (Figure 4-2). Improved forecasting helps utilities add only the capacity (transmission and distribution as well as generation) that is really needed, which holds down cost increases. Improved energy service planning allows better utilization of existing capacity, which reduces the fixed charge component of the rates. And improved customer relations help the utility to better understand and solve customers' problems, which increases the value of service to customers.

This chapter examines how refocusing on the customer helps improve forecasting, energy service planning, and customer relations, and discusses how utilities can increase their focus on customers.

Figure 4-2. IMPROVE SERVICES TO INCREASE CUSTOMER SATISFACTION

IMPROVED FORECASTING

In recent years the uncertainty surrounding the prediction of future demand seems to have increased. For example, Figure 4-3 shows a range of feasible forecasts of future demand by the North American Electricity Reliability Council.[2] Which one is most likely to occur?

In the face of such uncertainty, a number of utilities have found their conventional forecasting techniques unacceptable. More and more of them are turning to end-use forecasting. In end-use forecasting, a customer's electricity use is broken down by appliance or end use. For example, a residential customer's annual consumption of 10,000 kWh might consist of 5,000 kWh for water heating, 1,200 kWh for refrigeration, 1,200 kWh for clothes drying, 800 kWh for cooking, and the rest for miscellaneous uses.[3] Future consumption for the customer would be estimated by predicting the annual consumption for each appliance. End-use forecasting has proved particularly useful for incorporating the effects of conservation programs such as improved appliance efficiency and insulation.

Figure 4-3. SUMMER PEAK DEMAND PROJECTIONS

Comparison of Annual Ten-Year Forecasts
(contiguous U.S.)

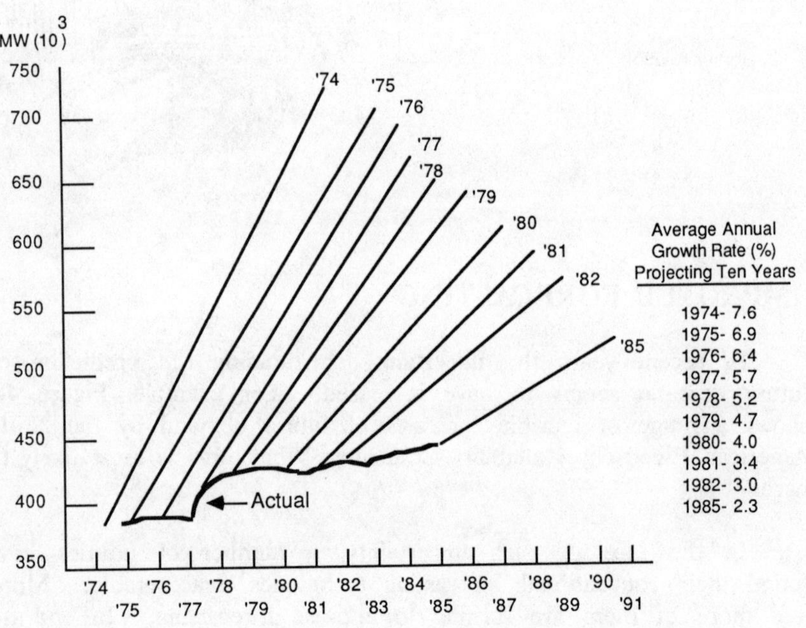

Source: North American Electricity Reliability Council[2]

This work with end-use forecasting has illustrated the importance of refocusing on customers to understand their preferences and behavior. For example, an energy efficient air conditioner does not itself choose to be located in a particular home; the electric range does not turn itself on. People choose and turn on appliances. Thus, a sound understanding of the customer's preferences and behavior is required to do a good job of end-use forecasting.

The importance of understanding the customer's preferences and behavior is illustrated by the results of some recent studies. Figure 4-4 shows the results of a study[4] on the preference or willingness of commercial customers to reduce their use of several end uses -- heating, cooling, water heating, and lighting -- in order to reduce their electricity bills; i.e., it shows the short-run and long-run price elasticities by various end uses. The figure shows that commercial customers have been more willing to reduce their heating and hot water use than their cooling or lighting use in order to reduce their electricity bills.

Figure 4-4. COMMERCIAL CUSTOMER WILLINGNESS TO CHANGE SO AS TO REDUCE BILL

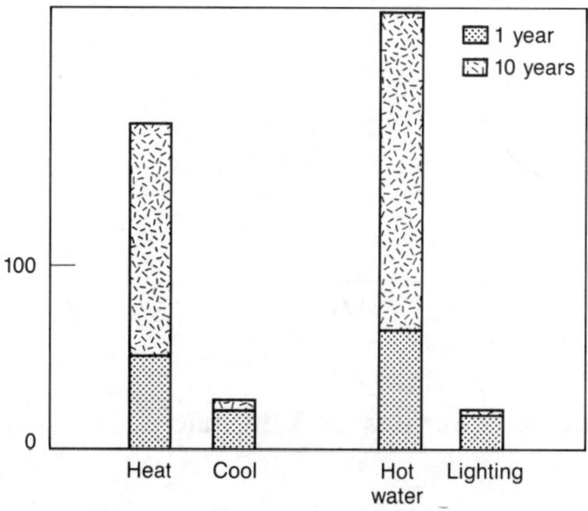

Source: Electric Power Research Institute[4]

The importance of understanding customer preferences is also illustrated by another study.[5] This study found that residential customer preference for electricity use in the 1970s was not really much different from that of the 1960s. After the differences in price and income growth rates between the decades are adjusted, similar growth rates in electricity use emerge (Figure 4-5). In particular, the actual growth rate of the late 1970s (3.3%) increases to 5.4%, which is much closer to the 7.7% experienced in the 1960s. Evidently, basic customer preferences have not changed as much as some people say.

Figure 4-5. REEPS ANALYSIS OF RESIDENTIAL ELECTRICITY DEMAND, 1975-1981

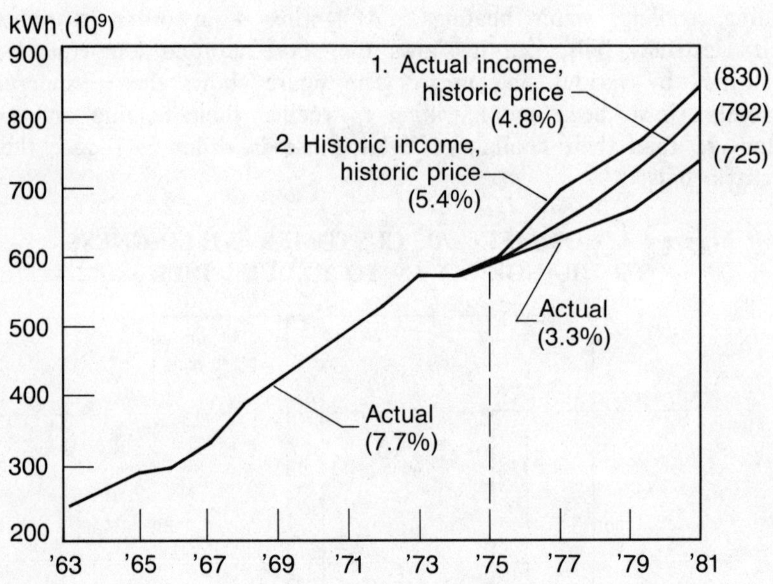

Source: Electric Power Research Institute[5]

It is clear that understanding customer behavior and preferences can help reduce the uncertainty in forecasting.

IMPROVED ENERGY SERVICE PLANNING

Understanding customer behavior and preferences can also help improve energy service planning. For example, utilities are increasingly aware of the fact that customers do not really want electricity itself; rather, they want the services it provides--warmth, hot water, work accomplished, etc. As a result, a number of utility organizations, such as the American Public Power Association, have broadened electricity forecasting and planning to include energy services planning.[6]

But how does a utility decide which energy services to provide its customers? The first and most important step is for the utility to determine what the customers want. Products and services offered must meet the customer's needs -- not the utility's. For example, utilities have to structure the product "off-peak power" in such a way that it meets customers' needs if the customers are going to buy it. The graveyard of U.S. businesses is filled with companies which had a "superior" product that met no one's needs.[7] The popular study of successful U.S. businesses, In Search of Excellence, underscores the importance of starting with the customer.[8]

A second critical step in determining which energy services to provide is to determine the utility's capabilities and needs. For example, if a utility currently (and for the foreseeable future) has tight reserve margins or is buying expensive purchased power, promoting sales does not meet the utility's needs. Instead, promoting the product "insulation" would probably help meet the utility's needs. As another example, most utilities don't possess the skills and resources necessary to manufacture heat pumps, even though heat pumps may be in the customers' best interests. Thus, manufacturing heat pumps is not a service utilities should provide.

The third step is to identify which utility energy services meet both the customer's and the utility's needs[9]; i.e., which services mutually benefit the utility and the customer. For the utility with tight reserve margins or expensive bills for purchased power, promoting insulation or other conservation measures would benefit both utility and customer. The utility would benefit from reduced use of expensive new power and the customer would benefit from lower rates.

A number of research projects are underway that can help utilities better meet their own and their customers' needs. For example, one project examines the effect of rates on promoting energy management programs.[10] Figure 4-6 shows some sample results of a simple method developed from research related to that project. These results are for a utility interested in promoting conservation and trying to choose between a declining block rate and an inverted block rate with a credit as the method by which to promote their goal. The results show that for all customers except the smallest users, the utility received almost all of the benefit from the declining block rate method of promoting the conservation program. In contrast, the inverted block rate with a credit method allowed the utility and customers of all sizes to share equitably in the benefits.

Figure 4-6. EFFECT OF RATES ON CONSERVATION BENEFITS

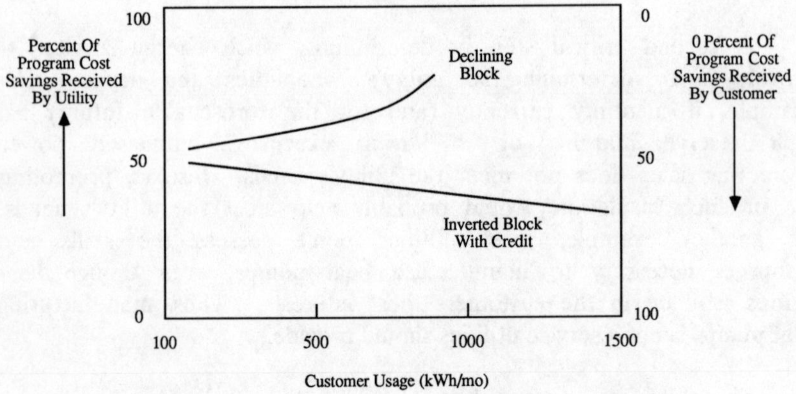

Source: Electric Power Research Institute[10]

Clearly, understanding and anticipating customer behavior and preferences is the cornerstone of energy services planning. The utility can be reasonably sure that customers will react favorably to its energy management programs if it has a good feel for their behavior patterns and preferences before these programs are finalized.

IMPROVING CUSTOMER RELATIONS

Understanding customer behavior and preferences is also important in improving customer relations. Unlike meters and appliances, the customer has feelings and perceptions that define reality for that customer. The utility must understand these feelings and perceptions.

Earlier it was noted that many customers believe they are not getting their money's worth from the utility. Put another way, the customers' perceived value of the utility's service is lower than their utility bills (Figure 4-7). It is not surprising that many utilities have been receiving increased numbers of customer complaints.

Figure 4-7. CUSTOMERS COMPLAIN WHEN THEY PERCEIVE THE COST OF ELECTRICITY TO BE HIGHER THAN THE VALUE

Strategically, utilities have two choices to bring the value and cost of service back into balance for the customer.[11] The first choice, as illustrated in Figure 4-8, is to reduce the cost of service below its value to the customer. This "least cost strategy" is the approach utilities have historically used -- find the least cost way to meet the customer's energy service needs.

Figure 4-8. UTILITIES HAVE HISTORICALLY FOCUSED ON REDUCING THE COST OF ELECTRICITY

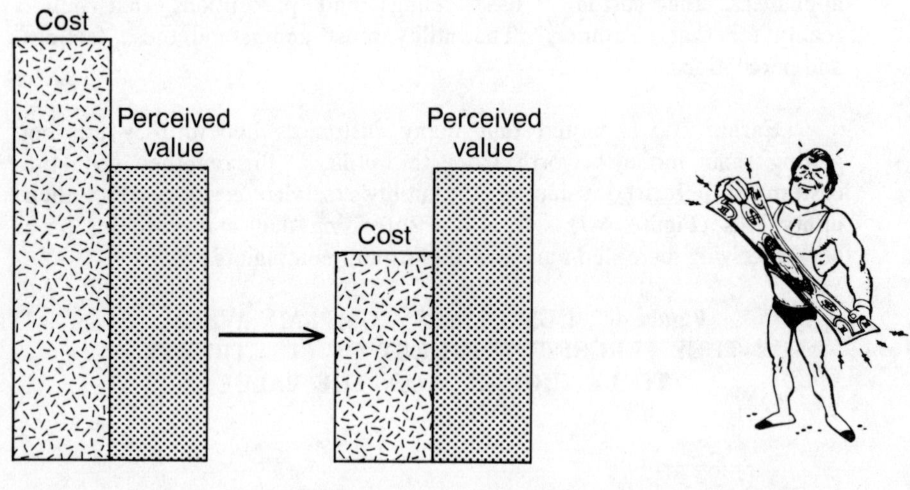

However, utilities with 40% reserve margins can cut costs only so far. For these utilities and others, a second strategy is desirable. This strategy (Figure 4-9), although foreign to utilities, is to increase the value of the service to the customer. In fact, In Search of Excellence reports that this second strategy is favored by "excellent" U.S. businesses. That is, these businesses tried to maintain customer satisfaction and their own financial health by increasing the value of their products rather than by minimizing costs. Presumably, this strategy should work for utilities. But how can a utility implement such a strategy?

Figure 4-9. RECENTLY, UTILITIES HAVE FOCUSED ON INCREASING THE VALUE OF ELECTRICITY

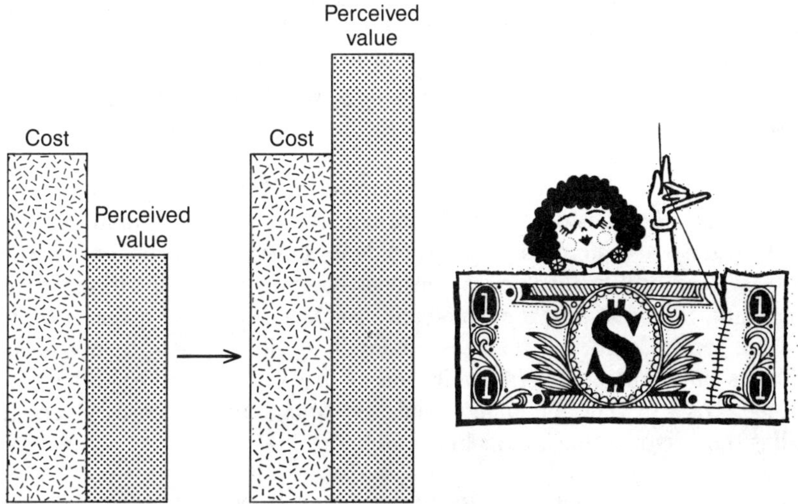

Utilities can increase the value of their service in several ways. First, they can change customer misperceptions. For example, Gulf Power Corporation was bracing for a furor from its customers over a pending rate increase. As background, they researched customers' attitudes and perceptions about their electric service. Surprisingly, Gulf Power learned that many of its customers mistakenly believed that the utility owned the gas wells from which it fueled its power plants. After an advertising/education campaign that corrected this misperception and others, there was a much smaller uproar over the rate increase than was originally anticipated.[12]

A second approach to increasing customers' perceived value of service is to provide a choice of services. A number of utilities have found that customers like having the choice between direct control of their central air conditioner (or water heater) and conventional service.[13] Yet, other surveys show that a number of customers prefer time-of-use rates to either direct control or conventional service.[14] Utilities could increase customer satisfaction (and the value of the utilities in the customers' eyes) by offering a choice of time-of-use rates. Perhaps, from a particular utility's perspective, time-of-use rates may not be a "least cost" alternative. But from some customers' perspectives, time-of-use rates provide greater value than the other alternatives -- and they are willing to pay for that value.

Finally, a third approach to increasing value is just plain old caring for the customers. Customer service representatives who really listen to customers and try to help solve their problems greatly increase the utility's value in the customers' eyes. In this day of increasing automation, the personal touch can mean a great deal. Indeed, one California bank is maintaining its human tellers along with its automated tellers, while other banks cut human tellers to reduce costs. The "human" bank is trying to increase value by selling the personal touch of its human tellers.

Figure 4-10. TYPICAL OUTAGE COSTS

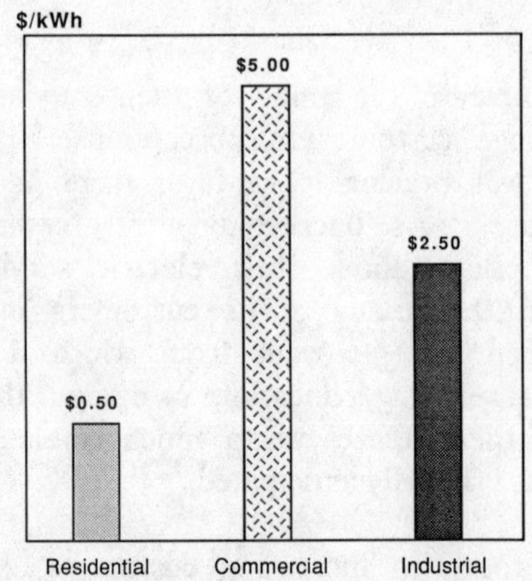

Source: Jack Faucett Associates[15]

To help utilities develop energy management programs that reflect the customers' value of service, studies are being conducted to evaluate the value of service.[15] One measure of value is the cost of a power outage (or, more generally, the value of that "lost" service) to various types of customers. As Figure 4-10 shows, outage costs are typically the highest for small-medium commercial customers and the lowest for residential customers. There is, however, considerable variation among customers within a class.

In sum, understanding customer behavior and preferences is important to improving customer relations, forecasting, and energy service planning. The next logical question is, "How do utilities obtain such data?"

OBTAINING CUSTOMER BEHAVIOR AND PREFERENCE DATA

To obtain data on customer behavior and preferences, utility planners must become cultural anthropologists -- or practitioners of some other social science. They can no longer think as traditional engineers do. Of course, this is an exaggeration, but consider the following two lessons learned from cultural anthropology and their implications for utility planners.

The first lesson is that to really obtain data on customer behavior and preferences, one must "get inside customers' heads."[16] That is, customers' needs/preferences and the resulting behavior are determined mentally as well as physically. Most people are familiar with physical needs such as food, clothing, and shelter. But mental needs such as security and sense of worth are not as familiar. And, significantly for utility planners, the way people choose to meet these mental needs can have a considerable effect on energy use. Thus, planners need to get inside customers' heads to understand and monitor these needs.

The importance of getting inside customers' heads is illustrated by the following familiar example. For many women in the 1970s, employment <u>increased</u> while raising children and homemaking decreased as ways to meet the mental need "sense of worth."[17] This shift set in motion a whole series of behavioral changes that have significant energy use implications for society. More young women are remaining single longer -- in 1970 about one-third of the women aged 20-24 were single, but by 1980 nearly one-half of women in that age bracket were single. The percentage of women of working age who were employed went from 40% in 1970 to 50% in 1980, and is projected to reach 60% in 1990. The fertility rate, or number of children born per woman, declined from 2.48 in 1970 to 1.86 in 1979.[18] These trends are leading to an older society with the number of households growing faster than the number of people. Older people spread among more households tend to use more energy than younger people gathered in the same household. One study reveals that this shift in demographics, caused particularly by the shift in women's attitudes, will lead to a 2% increase in energy consumption even if there is no growth in real income.[19] Thus, getting inside customers' heads to understand attitude changes is crucial if utility planners are to understand and influence energy use.

A second lesson learned from cultural anthropologists is that utility planners must be careful not to let their perspectives or world-views distort an understanding or interpretation of customers' perspectives or world-views.[16] Moreover, customers have a varying number of world-views -- and each one is essentially as valid as another. Furthermore, although he or she is a technical energy expert, the utility planner's perspective is not inherently any more valid than the customers' perspectives.

The importance of recognizing that the customer's world-view is different from, yet just as valid as, the utility's is illustrated by a University of Illinois study with young pigs.[20] The technical experts had always believed that young pigs grew best in a constant 70° Fahrenheit environment. In the university's study, a researcher taught the pigs to use a switch to control their own room temperature. The researcher found that the pigs did not prefer a constant 70° environment. Rather the pigs preferred varying temperatures -- often down to 60° at night. In fact, the temperature profile that the pigs preferred required 50% less energy for heating than the constant 70° selected by the experts. And these pigs grew as fast or faster than the pigs in the 70° environment.

If technical experts are unable to determine the best energy service solution for pigs, how much less are utility technical experts able to figure out the best energy service solution for utility customers? Perhaps one implication of this question is that more than one cost-effectiveness or cost-benefit calculation should be performed to evaluate programs. Each customer has a slightly different world-view, preferences, and value-tradeoffs. Each customer has a somewhat different cost-benefit calculation and preferred energy solution; thus the importance of giving our customers a choice of options.

Assuming that utility planners learn these two lessons from cultural anthropologists -- get inside the customer's head and recognize the validity of the various customer world-views -- how do they obtain customer behavior and preference data? Is it necessary to interview all customers? That is a possibility, and doing so can be a very important source of behavior and preference data. But there are two other sources which should be checked first to reduce data collection costs: The Census and other utilities' experiences. Let us consider each source.

As most people are aware, there are some excellent sources of existing data on people and their behavior. The Bureau of the Census and Bureau of Economic Analysis in the Department of Commerce and the Bureau of Labor Statistics in the Department of Labor are excellent federal government sources. The behavioral changes cited above relative to the shift in women's attitudes toward working came from the Bureau of the Census. Most states and cities have departments with tax, licensing, or budgeting authority that collect similar data.

There are also existing sources for attitude and preference data, such as academic social science literature. Market research firms systematically monitor attitude trends. The results gathered by some firms, such as the Gallup and Harris Polls, are available at no cost in the public domain. The results put together by other firms such as Yankelovich, Skelly & White/Clancy, Shulman are more comprehensive but cost thousands of dollars.

Another source for preference and behavior data is other utilities. In fact, some recent research shows that customers may be more similar across utilities than previously thought. For example, Figure 4-11 shows that the relative importance of the initial installed costs versus the operating costs in determining customers' preferences for electric heat is about the same in the Pacific Northwest as in the rest of the United States.[21] Presumably, any other utility could apply this customer preference information to the projected installation and operating costs of electricity and other heating fuels in its service area to determine the percentage of people preferring electric heat at some point in the future.

The transferability of customer data from one utility to another is also illustrated by another study.[22] Figure 4-12 illustrates the response or willingness to change electricity use based on 14 different residential time-of-use rate tests in five regions of the country. It shows that, after correcting for the differences in the ratio between peak and off-peak prices, the response of customers is almost the same. That is, with response (or elasticity) having plausible values ranging from 0 to 2.00, all the tests had response values ranging from 0.10 to 0.20, with 0.13 being the average. Thus, customer preference and behavior data seem more transferable among utilities than was previously thought.

The third way of obtaining customer preference and behavior information is for utilities to collect it individually. One existing source of customer behavior data is billing records. Most public utilities already have a source of customer preference data -- the results of elections of public utility council or board members and the transcripts of hearings on rate increases. Some utilities may want to supplement this information with other data sources -- surveys, focus groups, customer advisory panels, public workshops, etc. Because this customer research is so new to most utilities, monographs on "managing customer research," "market analysis," and "attitude/image measurement" are available to utility managers.[23]

Figure 4-11. PREFERENCES FOR ELECTRIC HEAT

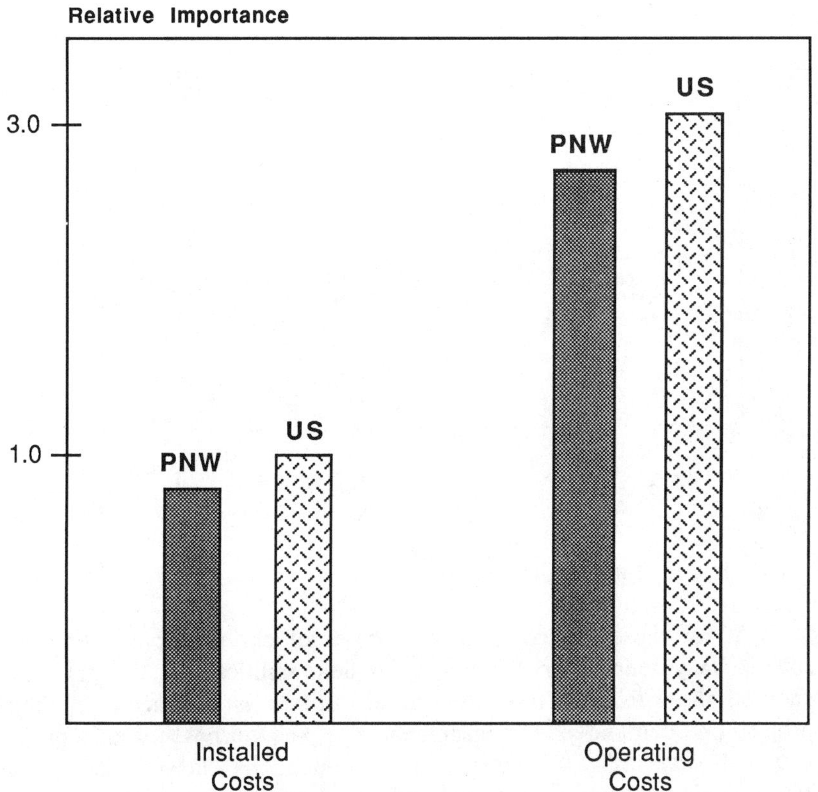

Source: Cambridge Systematics, Inc.[21]

Operating costs were roughly three times as important as installed costs in influencing customers' choice of electric heat in the Pacific Northwest region, as well as in the entire U.S..

Figure 4-12. RESPONSE TO TIME-OF-USE RATES

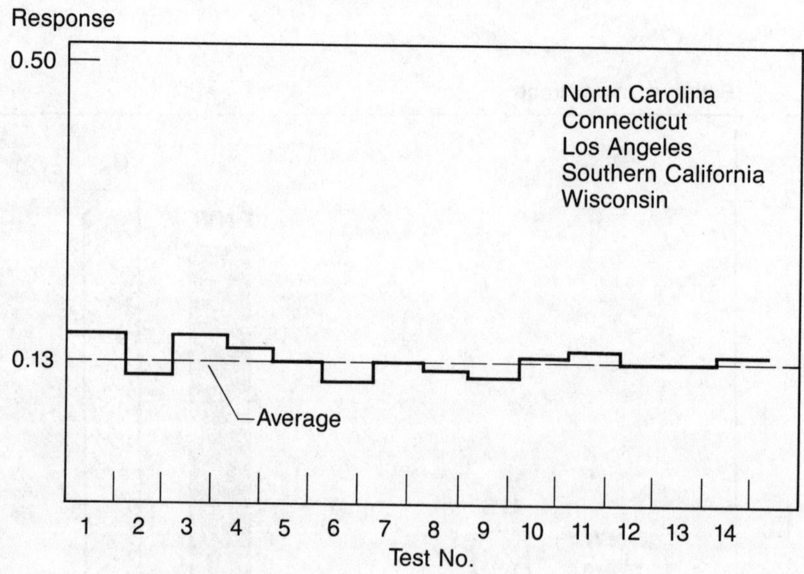

Source: Douglas Caves et al.[22]

All of these choices regarding types of energy services and data sources may seem overwhelming. To help utilities sort through these choices, the Electric Power Research Institute is conducting a major project on "demand-side management."[24] The purpose of this project is to provide utility planners with a guidebook that will help them select and implement the appropriate demand-side management or energy service programs. One key feature of this guidebook is that it will integrate all the current data on demand-side management programs into an easily usable form. The project has already revealed that a number of utilities will probably not need sophisticated analyses to select appropriate demand-side management programs. As Figure 4-13 illustrates, there are various levels of evaluation. The first two levels, intuitive and preliminary screening, require only simple methods, and for many utilities these two levels are sufficient to select the appropriate programs.

Figure 4-13. LEVELS OF EVALUATION

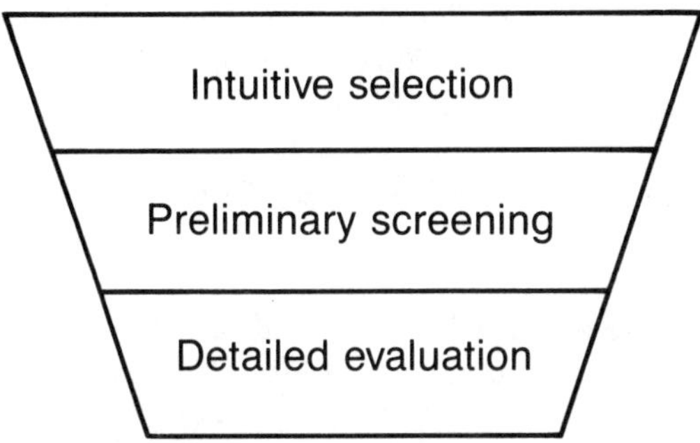

Source: Battelle-Columbus Division[24]

SUMMARY

Customers want their money's worth from electric service. The key solution is for utility managers to refocus on the customers. To this end, utility managers need to become cultural anthropologists and "get inside customers' heads." By better understanding customers' energy service needs, preferences, and behavior, utility managers can improve forecasting, energy services planning, and customer relations. These improvements will lead to a more efficient use of the utility's resources and more effective targeting of services. Perhaps then the customers will feel that they are getting their money's worth from their utility.

REFERENCES

[1]Cambridge Reports, Inc. "American Attitudes Toward Energy Issues and the Electric Utility Industry." Prepared for the Edison Electric Institute, May 1982.

[2]North American Electricity Reliability Council. "North American Electricity Reliability Council Annual Report 1981," p. 7.

[3]Electric Power Research Institute (EPRI). Patterns of Energy Use by Electrical Appliance (RP 576). Electric Appliance Energy Consumption Survey (RP 576-2). Commercial End-Use Model (RP 1216). Commercial End-Use Survey Design (RP 1216-4). Residential End-Use Planning System (RP 1918).

[4]Georgia Institute of Technology and Criterion, Inc. Commercial End-Use Model. EPRI RP 1216.

[5]Braithwait, Stephen, EPRI. Internal analysis using the model developed from the Residential End-Use Planning System (RP 1918). August 1983.

[6]Radin, Alex. "Energy Services Planning." Public Power 39, 3. May-June 1981, pp. 6-10.

[7]Levitt, Theodore. "Marketing Myopia." Harvard Business Review 48, July-August 1960, pp. 45-56.

[8]Peters, Thomas J. and Robert H. Waterman, Jr. In Search of Excellence: Lessons from America's Best Run Companies. New York: Harper & Row, 1982.

[9]Faruqi, Ahmad, EPRI. "Demand-Side Management: RPA2548." Presentation to the EPRI Energy Analysis Task Force, Orlando, Florida, August 23-25, 1983.

[10]Chamberlin, John, EPRI. Internal work on "Impact of Rate Structure on Energy Management Programs." RP 2440, August, 1983.

[11]Flory, John E. "Reflecting Customer Needs in Demand Side Management Planning." EPRI TPS82-652, June 17, 1983.

[12]Klein, Frederick C. "Marketing: Researcher Probes Consumers Using Anthropological Skills." The Wall Street Journal, July 7, 1983, p. 23.

[13]Thomas A. Heberlein and Associates. "Customer Acceptance of Direct Load Controls: Residential Water Heating and Air Conditioning." EPRI EA-2152, December 1981.

[14]Elrick & Lavidge, Inc. "Attitudes and Opinions of Electric Utility Customers Toward Peak-Load Conditions and Time-of-Day Pricing." EPRI Electric Utility Rate Design Study. January 3, 1977. And John E. Flory et al. "What Level of Electricity Service Do Californians Want?" Presented at the Second International Conference on Energy Use Management, Los Angeles, October 22-26, 1979.

[15]Jack Faucett Associates. "Power Shortage Costs: Estimates and Applications -- Volume 3." EPRI EA-1215, December 1981.

[16]Hunter, Yvonne L, Cultural Anthropologist and Director of the Energy Resources Management Assistance Program for the League of California Cities. Personal Communication, August 9, 1983.

[17]DeBoer, Connie. "The Polls: Marriage -- A Declining Institution?" The Public Opinion Quarterly, 45, Summer 1981, pp. 265-275.

[18]Barabba, Vincent P., Director, Bureau of the Census. "Demographics That Will Shape the Next Decade" in Attitude Research Enters the 80's, Proceedings of the American Marketing Association's Attitude Research Conference, Carlsbad, California, March 2-5, 1980.

[19]Angel Economics Reports. "Residential Energy Requirements with Zero Income Growth Expectations." Forthcoming report, EPRI TPS82-656-1.

[20]Galvin, Cindy. "Piglets Don't Hog Heat, Chop Nursery Bills 53%." Energy User News, 8, 33. August 15, 1983, p. 6.

[21]Cambridge Systematics, Inc. "Appliance System and Fuel Choice: An Empirical Analysis of Household Investment Decisions. Draft Final Report." EPRI RP 1918-1, May 4, 1983.

[22]Caves, Douglas W., et al. "Transferability of Customer Response to Residential Time-of-Use Electricity Prices." Presented at the Annual Review of Demand and Conservation Research, EPRI RP 1956-1, May 18, 1983.

[23]Hendrix, Philip E. "The Application of Consumer Research Techniques to Electric Utilities: A Scoping Study." Prepared by EPRI, April 7, 1983.

[24]Battelle-Columbus Division. "Scoping Study on Demand Side Management, Draft Final Report." EPRI RP 2381, August 1983.

CHAPTER 5

Understanding Customer Decisions: The Multiple Decision-Maker Problem

Dilip R. Limaye
Laurel Andrews
Craig McDonald

PROBLEM DEFINITION

Need to Understand Customer Decision Processes

Understanding the customer decision-making process is a critical element in the assessment of potential customer participation in utility marketing programs. Utilities are increasingly interested in the identification and assessment of the major factors that influence customer decisions; taking such factors into account enables the utility to increase customer participation in utility programs that are designed to provide benefits to both the customer and the utility. Understanding customer decisions requires information on customer preferences and behavior, such as:

- How do customers make decisions regarding purchase of energy-using equipment or appliances?
- What factors influence these decisions?
- How can utilities develop more effective marketing programs that lead to significant changes in customer behavior?
- How can such programs be effectively targeted to the appropriate customer segments?

- What are the characteristics of customers who are most likely to participate in specific programs, or to undertake specific behavioral changes?

To answer these types of questions, it is necessary to:

- Identify the decision-maker(s)
- Define the key product/service attributes and their influence on the decision
- Identify the acceptable and desirable levels of attributes
- Develop information on customer needs and benefits
- Define customer attitudes, preferences, and trade-offs relative to utility product and program attributes.

Multiple Decision-Makers

A complication that is commonly faced by researchers interested in understanding and defining customer preference and behavior is the role of multiple individuals and/or multiple entities/organizations in the decision-making process.[1] When such multiple decision-makers are involved, their roles and responsibilities in the decision process, and the major factors that influence their contributions to the final decision, are difficult to establish and represent in a manner helpful to utilities. However, if multiple decision-makers are not appropriately identified and represented in the analysis and modeling of customer preference and behavior, the results may be misleading. A major problem faced by utility planners is therefore how to understand and take advantage of the joint dynamics of multiple individuals or multiple entities in the process of decision-making regarding purchases of end-use equipment or appliances, or acceptance of utility programs and initiatives.

The problem of multiple decision-makers is common to a number of industries that make use of marketing and market research. In traditional consumer markets, the changing and fluid roles of men and women have made most product categories subject to multiple decision-making in two-person or larger households. An example is the automobile, which has traditionally been marketed to men, but which is now being targeted to women as well.[2,3] The problem is further complicated in categories where children are influential in the consumer decision, such as choice of fast-food restaurants.[4] In commercial and industrial markets, multiple decision-making is the rule, rather than the exception.[5] Cost-conscious companies are examining their purchasing more closely than ever, while technology has given us products which cross departmental boundaries and which have resulted in new and more involved buying decision processes.

Some Examples of Multiple Decision-Makers in Energy End-Use Markets

There are numerous examples of the multiple decision-maker problem as it relates to decisions that influence energy and/or electricity utilization. In each of the major end-use markets--residential, commercial, and industrial -- several individuals or entities shape the final decision. For example, in the residential sector, decisions regarding the purchase of many multifeature appliances involve both husband and wife, with one or the other taking a lead role in the decision process for specific appliances.[6,7] This is particularly true for major appliances such as refrigerators, air conditioners, freezers, etc. Other residential energy use decisions, particularly some that influence the pattern of utilization of electrical appliances, may involve not only the husband and wife but also teenage children living in the household. The influence of specific individuals in the decision process will generally vary from one household to another, depending on customer demographics, housing characteristics, appliance mix, and other factors including attitudes and preferences.

In the residential new construction market, decisions regarding choice of heating and cooling fuel and equipment as well as the selection of major appliances such as water heaters, dishwashers, air conditioners, etc., are generally made by the homebuilder or the contractor rather than by the homeowner. The homebuilder often bases his equipment selection decision on his perception of what will appeal to the homebuyer, but this decision is also heavily influenced by the builder's cash flow, cost of capital, and other factors.[8] The interaction between the builders and the contractors may vary significantly, with builders making decisions on some occasions and contractors on other occasions.

In commercial buildings, the process of decision-making regarding energy equipment and utilization patterns is influenced by an even larger number of different types of decision-makers. For example, decisions regarding the selection of fuel and equipment for heating and cooling of new commercial buildings may be influenced significantly by architect/engineers who are retained by the owners and developers of buildings.[9] The criteria used by owner-occupants, builders, developers, and architect/engineers to select among available equipment and fuel choices may vary substantially. Moreover, the specific influence of architects and engineers in selecting equipment and fuel may also vary depending on the type of owner or developer, the type of building in question, and other factors such as availability of specific programs or incentives from local utilities, and whether the owner/developer intends to occupy or retain the building.

With respect to decisions influencing energy use patterns in existing buildings, building owners, building managers, and tenants all participate in various aspects of the decision-making process.[10] Again, depending on the type of decision, the type of building, and the relationship between the tenant, the building manager, and the building owner, the influence of each party and the factors that influence these decisions may vary significantly.

In industrial settings, there may be several levels of hierarchy in the decision process; plant engineers, plant managers, financial executives, corporate staff, and corporate executives may all become involved in various aspects of the decision. The complexity of the decision process and the number of different decision-makers involved will be greater for large organizations and for decisions involving larger amounts of capital and/or operating expenses, such as in the lumber and wood products industry.[11]

The utility planner interested in influencing customer decisions regarding choice of fuel and equipment or equipment utilization patterns is therefore faced with the challenge of understanding the decision-making process, the role of various decision-makers in this process, and the factors influencing each decision-maker in order to design and target programs that will gain widespread customer acceptance.

KEY QUESTIONS FACED BY THE UTILITY MARKET PLANNER

With respect to the multiple decision-maker problem faced by the electric utility market planner, a number of key questions need to be addressed. Some of these are listed below:

- Who actually makes the decisions regarding equipment/fuel choice or energy utilization patterns? If multiple individuals are involved, who is the final decision-maker?

- What is the decision-making process and how can it best be characterized?

- Do the decision-making processes and the roles of the different decision-makers vary in different situations -- across different customer types or decision situations, or even within a specific customer type or decision situation?

- How frequently do different decision-makers make the final decision regarding fuel/equipment or utilization?

- Who provides inputs to the decision-making process?

- What types of inputs are provided? For example, the inputs to the decision-making process may involve a screening of potential choices, advice regarding the characteristics or benefits of specific options, evaluation of alternatives, and/or recommendation of specific options.

- How does the decision-maker handle inputs provided by multiple individuals?

- What weights are placed on the different inputs by the decision-maker, and what process is used to translate these inputs into the final decision?

Electric utility market researchers or planners have not studied the multiple decision-maker problems sufficiently to answer these questions for residential, commercial, or industrial markets. Very little specific research has been funded by the utility industry to develop information that addresses these key questions. However, the information that is available from existing utility focus groups and surveys[8,10,12,13] indicates that:

- The decision process varies across different customer types.

- The decision process also varies based on the end-use technologies being considered.

- The decision process may be influenced by utility programs or incentives, but the specific influence of such utility actions may vary across customer and technology types.

- The complexity of the decision process is influenced by the magnitude of the total capital and/or operating costs involved and the degree of perceived risk to the various decision-makers.

- The factors influencing the decision-making process are quite different across different types of decision-makers.

- The relative importance or the weights placed on the factors influencing decisions are different across different decision-makers, and may also vary across different types of decisions for the same decision-maker.

- The decision-making process and the factors and weights influencing that process are not easily identifiable.

REVIEW OF RELEVANT MARKETING LITERATURE

Multiple decision-makers pose many challenges to the market researcher, including how to design a proper survey, what questions to ask of various decision-makers, and how to combine their answers in a way that best addresses the marketing issues. A review of the literature relevant to multiple decision-makers is provided below, first with respect to certain residential decisions and then with respect to organizational decisions.

Multiple Decision-Makers in the Household

Understanding, explaining, and predicting purchase behavior by multiple decision-makers in a household have been recognized as complex and challenging problems by market researchers and planners. Information on the decision-makers, decision processes, and decision criteria is important in market analysis and planning, particularly with respect to selection of the appropriate promotional media and message. Davis[14] studied the problem of measuring the relative influence of husband and wife in consumer purchase decisions. He found that a number of marketing and sociological studies[15,16,17] had relied on obtaining information from the wife, concluding that responses of wives could sufficiently describe the family's decision process. Some other researchers have provided evidence that, on an aggregate basis, the responses of husbands and wives to purchase influences are similar.[17,18,19] However, many other studies of household decision-making point out the differences between husbands and wives relative to who makes purchase decisions and what factors influence such decisions.[20,21,22]

Granbois[6] provides an excellent discussion of household decision processes for durable goods, including appliances. He reviews the decision process in terms of the four major stages -- problem recognition, search and deliberation, selection and outcome, and post-purchase behavior -- and examines the roles and influences of different family members in each stage. Some of the findings reported by Granbois and other researchers[23] are:

- The greater the cost of a product, the greater the tendency for two or more family members to be involved in the decision process.
- The lower the family income, the greater the tendency for two or more family members to be involved.
- The participation by different family members in a product purchase decision is likely to vary according to the degree of their direct use of the product.
- Specialization in purchase decision-making will increase over the stages in the family life cycle.
- Husbands and wives who contribute nearly equal resources to the family are more likely to jointly participate in purchase decisions.
- The roles of husbands and wives will vary over the different stages of the decision process.
- The roles and influences of husbands and wives will vary from one product to another.

Belch, et al[4] have examined the relative influences of husband, wife, and teenage children in household decision-making relative to five different products. The wife was found to have at least a moderate influence in all purchase decisions. The husband's influence was the greatest for television and automobile purchases. Teenage children had little influence on most decisions but did contribute in the choice of autos and food products.

Pollay[24] has postulated a model of family decision-making which addresses the multiple decision-makers. In this model the outcome of a family purchase decision is a function of factors determining the net utility of product alternatives to each family member. In the long run, family members are assumed to receive utilities proportional to their respective priority indexes. In the short run, Pollay postulates, decisions are made through a process of negotiation and cross-influencing, which can be quite complex, and the individual purchase decisions may deviate from the long-term priority-based solution. Pollay's model, while intuitively appealing, is extremely difficult to empirically test or calibrate.

In summary, the participation and influence of multiple family members in purchase decision-making is difficult to conceptualize and measure. Most of the literature reviewed suggests, however, that both husbands and wives have important roles in the purchase decision.

Multiple Decision-Makers in the Organizational Setting

The commercial/industrial* marketing literature recognizes that multiple decision-makers are quite common in an organization. There is a substantial body of literature on organizational buying behavior,[23] and several excellent reviews of the state of the art.[25,26,27] Cox[5] also provides a detailed discussion of the roles and responsibilities of different decision-makers in the industrial purchase decision.

A conceptual model of organizational purchase behavior was presented by Webster and Wind.[28] This model is illustrated in Figure 5-1. In this model, four sets of factors affect the buying decision process -- environmental, organizational, interpersonal, and individual.

---

*The marketing literature addresses the "commercial" and "industrial" sectors, as defined by utilities, under the general term "industrial."

Figure 5-1. CONCEPTUAL MODEL OF ORGANIZATIONAL PURCHASE BEHAVIOR

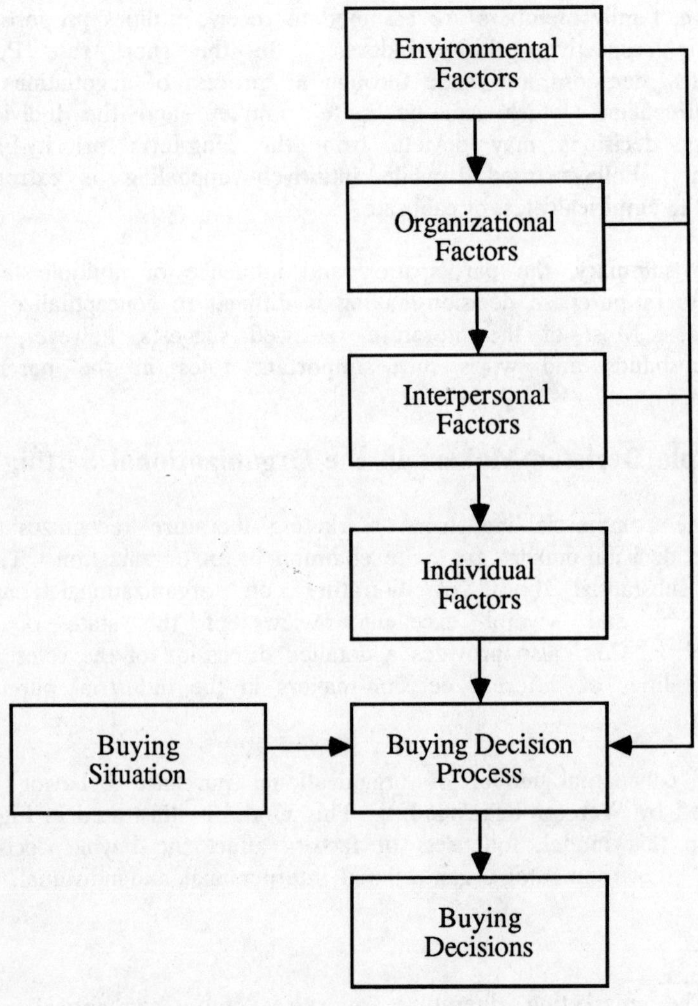

Source: Webster and Wind[28]

Each of these variables has two subcategories:

- **Task variables** -- Those directly related to the buying problem
- **Nontask variables** -- Those extending beyond the buying problem.

Environmental factors determine the availability of goods and services, the general conditions in which buying takes place, and the values and norms for individual and organizational behavior.[5] These factors include physical, technological, economic, political, legal, and cultural factors.

Purchase decisions take place within an organization and are shaped by a number of organizational factors, including the organization's purposes, objectives, and constraints. Organizations are composed of four sets of interacting variables: tasks, structure, technology, and people. Most research on the relationship between organizational factors and purchase behavior has focused on:

- **Structure** -- Impact of centralized/decentralized structures
- **People** -- "Buying center" membership and responsibilities.

The Buying Center Concept

The concept of the "buying center" was first introduced by Webster and Wind,[28] who defined it as composed of all individuals involved in the decision process. Interpersonal and individual factors related to members of the buying center significantly shape the purchase decision. Five different types of roles have been identified as being assumed by members of the buying center; it is possible for multiple individuals to play one role or for one person to play multiple roles. The five roles are:

- **Users** -- Members of the organization who will use the product

- **Influencers** -- Members of the organization who directly or indirectly influence the purchase decision

- **Buyers** -- Members of the organization who make the actual purchase

- **Deciders** -- Members of the organization who make the final purchase decision

- **Gatekeepers** -- Members of the organization who control the flow of information.

The involvement of persons "playing" each role varies over the five stages of the buying process:

- Identification of need

- Establishment of specifications

- Identification of buying alternatives

- Evaluation of alternatives

- Selection of the supplier(s).

Cox[5] notes that it is difficult to develop a methodology to accurately measure the role influence. He also cites studies showing that buying center members inflate their own influence relative to others' perceptions of their influence.

Kotler[29] points out that the purchase decision process and the roles played by different individuals in this process are also affected by the type of purchase situation, referred to as buyclass. He describes three situations:

- **Straight rebuy** -- A routine buying situation that calls for routinized response behavior. This type of purchase may be handled by a single individual.

- **Modified rebuy** -- Buyer modifies something purchased in the past. This involves limited problem solving and could be handled by a small middle management committee.

- **New task** -- New, unfamiliar offer from an unfamiliar seller. This may require extensive problem solving and is often handled by top management, including executives with different areas of expertise.

Robinson and Farris[30] also provide a descriptive model of industrial purchase behavior which characterizes the decision process according to purchase situation.

Nicosia and Wind[26] discuss the issues related to the selection of the unit of analysis to be used when conducting research on organizational purchase behavior. The four alternatives are: the individual, the buying center, the entire organization, and the aggregate market. The selection of the analysis unit depends, according to them, on the product and the buying situation. Cox[5] uses a different scheme for classifying the unit of analysis:

- **Individual unit** -- Person, department, or organization as a whole

- **Multiple unit** -- Two or more individual units

- **Interactional unit** -- Relationship between two or more individual units.

Modeling the Organizational Purchase Decision

Choffray and Lilien[31] present a general structure of organizational buying behavior (Figure 5-2). This structure addresses the following issues:

- Potential customer organizations differ in their "need specification dimensions" -- i.e., in the dimensions they use to define their requirements. They also differ in their specific requirements along these dimensions.

- Potential customer organizations differ in the composition of their buying centers -- in the number of individuals involved, their specific responsibilities, and in the way they interact.

- Decision participants, or individual members of the buying center, differ in their sources of information as well as in the number and nature of the evaluation criteria used to assess product alternatives.

Sheth[32] also provides an overall model structure for industrial or organizational buyer behavior. Figure 5-3 illustrates Sheth's model, which is designed to be a generic model that attempts to describe and explain all types of industrial purchase decisions. A number of the elements in Figure 5-3 can be simplified in order to analyze a specific product/purchase decision. Sheth's framework explicitly addresses the issue of multiple decision-makers, particularly in the gathering, evaluation, and assimilation of information, and the negotiation and conflict resolution process among individuals with different perceptions, needs, preferences, and goals.

All of the approaches listed above point out the need for understanding the composition of the buying center, the characteristics, needs, and preferences of the individuals, the interpersonal relationships, and the group cohesiveness. These types of information are extremely difficult to obtain in practice, particularly because of the diversity of industrial organization.

Figure 5-2. MAJOR ELEMENTS OF ORGANIZATIONAL BUYING BEHAVIOR

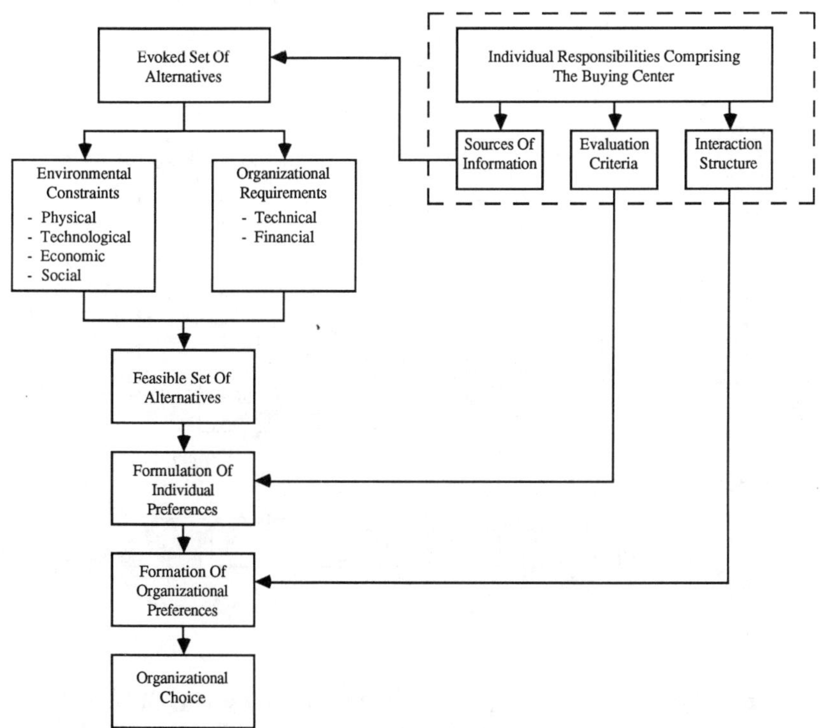

Wind and Thomas[27] discuss some of the complexities of analyzing buying center purchase decisions. They point out that:

- Little is known about the composition of the buying center, the determinants of its composition, or the influence patterns among members.

Figure 5-3. AN INTEGRATIVE MODEL OF INDUSTRIAL BUYER BEHAVIOR

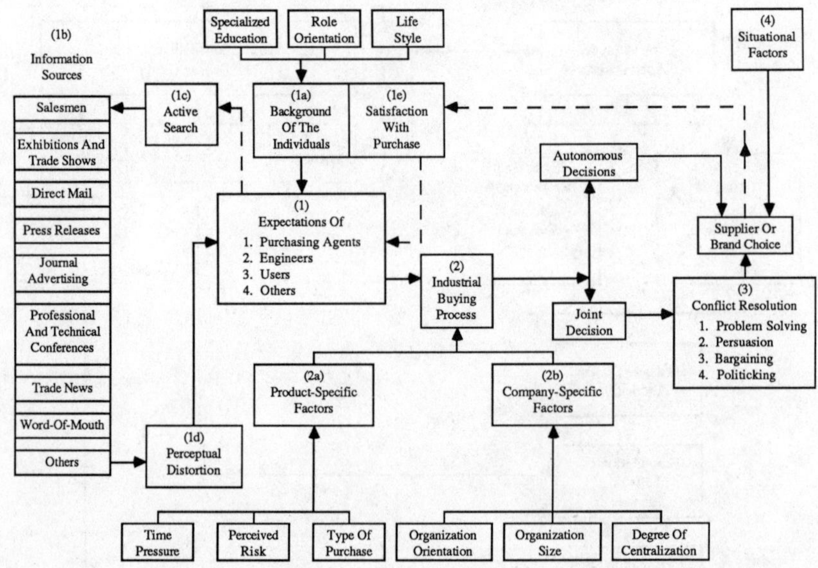

Source: Sheth[32]

- The literature suggests that the composition of the buying center varies by organization, and within organizations by situation and for idiosyncratic reasons.

- Illustrative role studies imply that members in the buying centers vary in their perceptions.

- Studies on influence are sensitive to the measurement approach used.

For example, studies using an attributional approach found expert power to be very important, while studies using an experimental approach found other bases of power to be more important.

Numerous other analytical frameworks have been proposed in the literature on organizational purchase decisions. Wind[33] describes a reward/measurement model based on the expectancy theory of motivation; i.e., that buying behavior is influenced by motivation and rewards and is a form of work behavior. Thus, purchase behavior is motivated by the reward system in a firm and by personal characteristics. Morris and Freedman[34] argue that coalitions within an organization influence purchase decisions. A number of researchers[35,36,37,38] contend that organizational purchase behavior can only be explained in the context of buyer-seller interactions.

Choffray and Lilien[39] present a set of four models for group choice behavior. These are:

- **The Weighted Probability Model** -- This model assumes that the firm as a whole is likely to adopt a given alternative among the feasible ones, proportional to the relative importance of those participants who choose it. A two-stage process is involved:

  -- Member of buying center is chosen as the decision-maker based on his relative importance in the choice process.

  -- Decision-maker makes choice.

  Conceptually this model assumes that a single decision-maker is responsible for the group choice.

- **Voting Model** -- This model assigns equal weight to all individuals in the buying center and assumes that the choice with the largest number of votes will be chosen.

- **Minimum Endorsement Model** -- This model assumes that, to be chosen, a product must be chosen by a quota of buying center members; i.e., participants vote over and over until a quota is achieved.

- **Preference Perturbation Model** -- This model assumes that the group will choose the model that perturbs individual preferences least.

A recent review of the theoretical literature on industrial purchase behavior by Johnston and Spekman[40] found that there were over a thousand references to articles and books on industrial marketing. However, according to these reviewers, "This research seems to have been insufficient in impact and failed to make much of an impression on marketing academicians and practitioners. The popular belief seems to be there is not a substantial amount of research or knowledge about industrial buying behavior."[40] Cox[5] also states, "The organizational buying process is complex, it may vary by product/industry and buying situation, it is difficult to model and most critical, it is difficult to validate empirically."

APPROACHES FOR ADDRESSING MULTIPLE DECISION-MAKERS

Marketing practitioners have developed information on criteria and factors influencing industrial decision-makers (e.g., through conjoint analysis), but many of the empirical studies are unpublished. The findings of some of the empirical research are summarized below.

Bellizzi[41] studied firms in the construction industry and found decision processes to vary by size, with large firms exhibiting a greater incidence of multiple decision-makers. Berkowitz[42] examined the real influencers in new product adoption in the industrial market. Choffray and Lilien[31] measured the influence of engineers, plant managers, financial officers, company executives, and consulting engineers in decisions to purchase air conditioning equipment.

Johnston[43] examined 62 industrial purchases and found that every single decision involved more than one decision-maker. Johnston and Bonoma[44,45] interviewed relevant people in organizations purchasing capital equipment and studied the composition of the buying center; they concluded that finding the "key buying influences" may be complex and difficult. Spekman and

Stern[46] interviewed 20 industrial firms to study the purchase decision-making process, and documented the structure of multiperson buying centers and the need for marketers to develop information on decision processes and criteria.

Today's commercial marketing research firms most frequently utilize one of three basic approaches to the multiple decision-maker problem. The simplest and most frequently used approach is the "key informant" method.[47] This entails identifying and selecting the individual who is most knowledgeable about the total decision-making process, and using a list of such individuals as the sampling frame. In the survey, the respondent is asked to answer on behalf of the entire decision-making unit. The dynamics of that unit are implicitly assumed to be reflected in the survey responses. Additional information pertaining to the interactions among the multiple decision-makers may be obtained from the point of view of the selected respondent category.

A second commonly used approach to multiple decision-making entails a pooling of data from several sources.[20,48] The major categories of decision influences are identified prior to the study, and samples of respondents are drawn from each category. The survey instruments are designed with a common core set of questions. This core usually consists of questions on attribute preferences, brand perceptions, customer needs, and problems. The core questions from each sample category are then pooled using some sort of weighting; e.g., in purchase decisions involving commercial HVAC there may be three sample categories -- owners, building managers, and architects/engineers -- whose answers must then be weighted. The weights used in pooling may be judgmentally derived, or they may be developed from respondents' assessments of the relative importance of each decision influence.[46,49]

The third approach to multiple decision-maker analysis is to create a functional model of the decision process which provides a simulation of the actual dynamics of that process and how they relate to market behavior.[50] Some examples follow:*

- **Consensus Management** -- Decision-making in many large organizations can be characterized as management by consensus.[51] In this process, alternatives are screened not on the basis of optimality, but rather on the basis of acceptability to all relevant parties. As a simple example, consider a four-person committee which has five alternatives for an important purchase. Suppose that these alternatives, labeled A through E, are ranked by each participant as follows:

	Participant			
Rank	Tom	Jane	Ed	Mary
1	B	A	B	B
2	C	D	D	C
3	A	C	C	A
4	D	E	E	D
5	E	B	A	E

  In a consensus formulation, the winner is C, which is nobody's first choice, but which ranks no worse than third to anyone. Alternative B, a majority first choice, is vetoed by Jane and therefore is not selected in this model -- but in a different model B would be selected.

*These examples are from proprietary studies conducted for manufacturers, and are based on personal communication with Morris Olitsky, President, MatheMarketing Inc., May 1986.

- **Selection from a Limited Choice Set** -- There are a number of purchase settings in which one party is the primary decision-maker, but the set of his choices is restricted by other parties, which are susceptible to different sets of marketing variables. For example, in a study of intravenously administered drugs used in a hospital, the attending physician was the primary decision-maker, but his choices of drugs were limited to those in an allowable formulary controlled by a hospital committee. The research included interviews of attending physicians and of chief pharmacists, the latter group representing the hospital formulary committees. A simulation model of the behavior of both groups -- attending MDs and formulary committees -- was developed, along with a model of their interrelationship. Marketing recommendations were developed in light of these dynamics.

- **Influential Third Parties** -- Many markets can be represented as consisting of a primary buyer whose choice is influenced by one or more other parties. Such situations may be found in purchases involving technical knowledge, such as personal computers or agricultural chemicals, where the supplier or dealer is often an important buying influence. The decision process was studied using personal interviews with samples of farmers and pesticide dealers. The farmers were asked to assess the importance of their dealers' recommendations relative to other attributes such as price. The significant market influences on both farmers and dealers were determined, and the importance of dealer influence in the decision was assessed in order to help develop the overall marketing strategy.

Many product categories are distinguished by having one person or group as the initial specifier of a purchase, and a different person or group as the consumer or user of the product. Both groups are important. Some examples are:

- **Pre-sweetened breakfast cereals** -- The parents buy it, their children (usually) eat it and provide feedback for the next purchase cycle.

- **Ethical (prescription) drugs** -- The physician writes the prescription, but the patient takes the medication as required, generally less often. The physician obtains some feedback on patient's compliance.

The three basic approaches to the multiple decision-maker problem each have their pros and cons.[52,53,54,55] The first approach mentioned, that of utilizing a single constituency to represent the decision-making process, has these advantages:

- Freedom from assumptions about who the decision influences are and how they interact

- Simplicity of design and execution

- Least cost and quickest turnaround.

Its disadvantages are:

- Difficulties in identifying the appropriate individuals

- Possible bias due to respondents' particular vantage point

- Difficulties in validation

- Limited understanding of the interactions among various decision influences.

The second approach mentioned, that of pooling responses from multiple sources, has these advantages:[51]

- Understanding similarities and differences among various points of view

- Explicit representation of all important purchase influences.

The disadvantages of the pooling approach are:

- Greater cost and complexity in the research
- Assumption that buying decisions can be represented by a pooling or averaging process.

Finally, the functional modeling approach has one major advantage -- its ability to represent the particular dynamics of a market. This strength is also a potential weakness, in that an inappropriate model, if selected, can produce an incorrect understanding of a market. A further disadvantage of the functional modeling approach, relative to the simpler single representative method, is greater complexity and higher cost of design and implementation.

CHARACTERIZING MULTIPLE DECISION PROCESSES RELEVANT TO UTILITY PROGRAMS

The multiple decision-making processes relevant to utility demand-side management (DSM) in the commercial and industrial sectors can be broadly classified into four generic types:

- Hierarchical
- Advisory
- Group
- Mixed.

Each of these is briefly discussed below.

Hierarchical

In this type of process, there are two or more levels of decision-making, with the final decision being made at the highest level. At the lower levels of decision-making, a screening process may take place and certain options may be screened out. Evaluation of alternatives may also be performed at the lower levels and, in some cases, a recommendation made to the final decision-maker. An example of the hierarchical decision process would be the relationship between the building manager and building owner in a commercial building. In such a situation, the building manager may evaluate alternatives and recommend to the building owner that he make a capital investment in an energy conservation or energy management device. The final decision, however, is made by the building owner. Similarly, in industrial situations the plant engineer, plant manager, and corporate executives may represent a three-level hierarchy with respect to decisions regarding energy-using equipment in the plant. The plant engineer may perform identification and screening of options, the plant manager may perform the evaluation and recommendation, and the corporate executive may be responsible for the final decision.

Advisory

In this type of process the final decision-maker receives various types of inputs from a number of different individuals or entities. The decision-maker may place different weights on these inputs. An example of this type of process in a commercial market is the role of architects and engineers in advising the developer or owner on the selection of fuels and equipment for heating and cooling. In some situations, the owner or the developer may get separate inputs from the architect, the mechanical engineer, and the contractor regarding the costs and efficiency of alternative systems before making the final decision on which particular system to install. Similarly, in the residential sector, the mechanical contractor may provide input to the homebuilder's decision regarding the size of and budget for specific equipment. Many examples also exist in the industrial sector. The plant manager or executive who has to make decisions regarding energy equipment may solicit advisory input from various departments such as engineering facilities, management, and production, and weigh these inputs differently before making the final decision.

Group

In the group decision-making process, a committee of decision-makers is involved; committee members vote on the specific choice of equipment, fuel, appliance, or other options. An example of this in the commercial sector is a syndicate or partnership of owners in which each member has equal say in the final decision. Similarly, in the industrial situation, a corporate committee involving representatives of many departments may make the selection, with each committee member having a vote in the selection process.

Mixed

A mixed decision process may involve a combination of two or more of the above processes. One example in the commercial sector is the situation in which the building manager screens and evaluates the options and presents the findings to a syndicate of owners, who then vote to make the final selection. In the industrial sector, an example would be a situation in which many individuals or departments provide inputs to the plant manager, who makes a recommendation to a corporate committee, which then votes on the final selection. Alternatively, there are situations in which a committee evaluates the options and a manager or executive makes the final decision.

ISSUES IN DATA COLLECTION AND ANALYSIS

The utility market planner faces a number of important and difficult issues in developing and analyzing information regarding the multiple decision-making problem in the residential, commercial, and industrial sectors. The first and most important difficulty is in developing an understanding of the decision-making process. In view of the different types of processes possible and the number of different decision-makers involved, it is not easy to represent a particular market or a particular decision in terms of a specific set of decision-makers using a single formal process. In fact, most cases will involve variations in the number and type of decision-makers and the specific processes involved.

The second major difficulty in developing adequate information is that there are significant variations in the decision criteria used by the different decision-makers and the factors that influence the decision-making process. Furthermore, the weight or importance of the different factors can also vary significantly. Such variations can occur across different decision-makers, different end-use technologies and programs, and different organization types. The differences may also be a function of the size of the organization and the complexity of the organizational structure.

Another major issue in adequately representing the decision process is the measurement of the knowledge and awareness of the different decision-makers relative to the programs or technologies of interest to the market planner. It has been found in many utility surveys, such as, for example, a recent survey to assess market penetration of cool storage conducted for Florida Power and Light Company,[13] that there are significant differences among different types of decision-makers representing different building types relative to the degree of knowledge and awareness regarding new end-use technologies. Such differences in knowledge and awareness make it more difficult to understand and represent the decision-making process.

There are a number of practical difficulties in collecting and analyzing information regarding the multiple decision-maker problem. In survey research, for example, key issues are how to identify the decision-makers and who to contact to obtain the relevant information regarding the factors influencing the decision process and the weights assigned to these factors. A related practical problem is that decision-makers may not necessarily be located at the customer's facility. This is particularly true with respect to commercial and industrial decisions; owners or executives are often found in a corporate headquarters facility far away from the customer's location in the utility's service area. Finally, even if the ultimate decision-maker can be located, it is difficult to identify all of the other decision-makers who provide inputs to the final decision, and to assess the factors and weights assigned to their inputs.

DATA COLLECTION METHODS

The methods used to identify the decision-makers, and to develop an understanding of the decision-making process when multiple decision-makers are involved, include focus groups, surveys, in-depth interviews, and conjoint data collection. Each of these is briefly discussed below.

Focus groups are small groups of individuals with similar interests. An in-depth discussion is conducted under the guidance of a moderator, who channels it into predefined areas of interest while remaining alert to the spontaneous comments that often provide insights into the research questions at hand. Focus groups provide a quick and relatively inexpensive means for gathering qualitative data regarding markets and decision-makers. With respect to the multiple decision-maker problem, focus groups can be used to identify the decision processes involved, the major decision-makers, and the factors and criteria influencing the decisions.

A number of utilities have used focus groups to identify the decision-making process relative to new commercial end-use technologies, and have found such groups to be useful for this purpose. Examples include focus groups, composed of architects, engineers, developers, building owners, and managers, used to explore the decision-making process relative to thermal energy storage installation in existing and new buildings -- conducted for Florida Power and Light Company[13] and North Carolina Alternative Energy Corporation[10] -- and to identify the factors influencing fuel and equipment choice in new commercial/industrial facilities -- conducted for the New England Power Pool[56] and Jersey Central Power and Light Company.[57] These types of focus groups have shown that utility planners can identify variations in the decision process across different organization types, as well as different program and technology types, using the focus group approach. Focus groups can also be used to help identify and even rank the significant factors that influence each decision-maker. However, focus groups cannot be used to quantify the specific weights that are placed by decision-makers on the different factors and criteria.

Various types of mail, phone, and on-site surveys have also been used to address the decision-making process relative to end-use technologies and programs. Mail surveys have been used by utilities to obtain information on customer characteristics. However, as far as the multiple decision-maker problem is concerned, mail surveys are generally of limited value because they do not allow for further questioning of the respondent relative to the decision-making process and the different individuals involved in that process. Phone surveys can be more effective ways to identify the decision-makers and the decision processes, but the information obtained from phone surveys is limited by the constraints on interview length imposed by the phone survey process. On-site surveys have been utilized to obtain more detailed information on decision-makers and decision processes. Among survey techniques, on-site surveys probably provide the best method for obtaining information regarding multiple decision-makers. However, they are also the most expensive. Examples of survey techniques used for addressing the multiple decision-maker problem include telephone surveys of architects/engineers, developers, and owner-occupants conducted by the New England Power Pool,[56] and surveys of industrial decision-makers conducted by Savannah Electric and Power Company.[58]

In-depth interviews involve detailed discussions with knowledgeable individuals conducted by experienced professional interviewers who are themselves knowledgeable regarding the major issues of interest. In-depth interviews can be conducted by phone, although the common method is to conduct them in person. Generally a detailed interview guide is prepared to channel the discussion into meaningful and relevant areas. Face-to-face in-depth interviews provide an excellent method for gathering information on decision processes and criteria. Examples of this method include interviews conducted with industrial decision-makers by Southern California Edison Company[59] and interviews with commercial decision-makers conducted by San Diego Gas and Electric[60] and El Paso Electric.[61] The successful use of in-depth interviews requires experienced technical interviewers, and may also require significant incentives to the potential respondents since interviews may last an hour or longer.

The final method for data collection is through conjoint exercises. Such methods can quantify the decision processes and criteria and have been utilized by Pacific Gas and Electric Company and Southern California Edison.[62] However, for conjoint methods to be successful in assessing multiple decision-maker situations, the decision process must be structured before the conjoint profiles are developed. Also, appropriate incentives must be provided to the respondents in order to ensure their participation and cooperation.

MODELING APPROACHES

Approaches to modeling the multiple decision-maker problem are still evolving. There is not a large body of literature on the subject, and there is no universally acceptable method for representing multiple decision-makers, either in the residential or in the commercial and industrial sectors. A brief overview of how multiple decision-makers have been handled in other industries was provided above. As far as the utility market planner is concerned, four alternative approaches for modeling multiple decision-makers are:

- Explicitly model each step in the decision process and the role of each different decision-maker
- Model the final decision-maker only

- Develop an organizational model

- Develop a weighting scheme for different decision-makers.

Each of these approaches is briefly reviewed below.

The first approach involves a simulation modeling framework in which each step in the decision-making process is explicitly represented and modeled. All of the different decision-makers are represented in such a framework, along with the criteria and factors used by each decision-maker and the weights assigned to each factor or criterion. This approach requires an excellent understanding of the entire decision-making process and the role of these decision-makers prior to development of the models. An advantage of this approach is its ability to represent the dynamics of the decision-making process and examine how different types of decision-makers and different decision processes will react to specific utility programs or incentives. The disadvantage is the complexity of the modeling process and the difficulties in developing appropriate data to represent the process. This approach might be suitable for a specific problem, but may become extremely cumbersome if a large range of alternative end-use technologies and programs is to be represented.

The second approach is to model the final decision-maker only. This method requires that information be developed on who the final decision-maker is, and what factors and criteria he uses to make the decision. The modeling process is somewhat simpler than for the first approach. However, this approach may not portray the decision-making process accurately since the inputs of other decision-makers and the screening and evaluation that take place at different decision-making levels are not represented.

The third approach involves developing a model of the organization as a whole as opposed to the individual decision-makers. In this approach information is gathered (through surveys) on the factors and criteria used by the organization as a whole -- though not necessarily by each individual involved in the decision process --

in making decisions. For certain types of commercial and industrial organizations, it may be possible to identify an individual within the organization who may be able to provide information from the entire organization's perspective. The potential advantage of this approach is that it may represent the dynamics of the multiple decision-maker problem without the complexity inherent in trying to represent the entire process in the model. The disadvantage is the need to identify the appropriate individual who can provide data on the organizational decision process and criteria rather than those of the individual decision-maker. Also, when decisions involve multiple organizations -- such as architect/engineers and developers or owners trying to select new commercial heating and cooling equipment -- this approach has limited value.

The fourth approach is to develop a model that provides information regarding the different decision-makers, and then to apply a scheme that weighs the decisions made by different decision-makers to come up with the final answer. This approach may be suitable for cases involving multiple organizations such as architects/engineers, developers, and builders, but appropriate weights need to be developed from survey research to represent instances in which each type of decision-maker makes pertinent decisions.

IMPLICATIONS FOR UTILITIES

This chapter has discussed the problems, challenges, and complexities of representing multiple decision-makers in addressing customer preference and behavior relative to end-use technologies and programs. It is clear that the utility must address multiple decision-maker issues in each of the three major sectors. Since there is a lack of significant research relative to this problem in both the utility and other fields, it is apparent that pioneering research is required.

With respect to the residential sector, multiple individuals within a single household may be addressed by obtaining information on the household decision criteria and weights. This may be accomplished by structuring household interviews or other data collection mechanisms such that appropriate information is developed on the different decision-makers as part of the data collection process. The more complex problem involves the new construction market, in which homebuilders and contractors become involved in the decision process. If such situations are to be analyzed, one of the approaches used for commercial or industrial situations may have to be utilized.

In the commercial and industrial sectors, since multiple decision-making is the rule rather than the exception, a method must be developed that better explains this process. There is a tremendous diversity of organizational structures, decision processes, technologies, and programs. This implies that there is also a tremendous diversity of decision processes, decision-makers, and factors and criteria that influence those decisions. The research conducted by utilities must therefore be structured and focused on well-defined and specific topics. The data collection and modeling approaches will have to be customized to the specific problem or decision being addressed.

It is suggested that utilities attempt to develop the organizational modeling approach -- representing the needs, preferences, and decision criteria for the organization or buying center as a whole, rather than for each individual decision-maker within the organization or the buying center. The data collection instruments should be structured to collect the data on organizational rather than individual perspectives. When different types of organizations are involved, the model should represent each type of organization and provide appropriate weights based on data collected through survey research.

The specific data collection methods and model structure will be dictated by the specific problems or decisions to be addressed. It is therefore recommended that the selection of the specific decisions of interest be made before embarking upon the data collection and modeling activities.

REFERENCES

[1]Electric Power Research Institute (EPRI), Project Description, Customer Preference and Behavior Project, Draft Report, Palo Alto, CA, May 1986.

[2]Brown, George H., "The Automobile Buying Decision Within the Family" in Nelson N. Foote (ed.), Household Decision-Making Consumer Behavior Volume IV, New York: New York University Press, 1981. pp. 193-199.

[3]Gay, Verne and Paul L. Edwards, "Chevy Woos Women -- Sports Dollars Cut for Major Effort," Advertising Age, 9-9-85, pp. 1, 130

[4]Belch, G. and M. Belch and G. Ceresino, "Parental and Teenage Child Influences in Family Decision Making," Journal of Business Research, 13, April 1985, pp. 163-176.

[5]Cox. W. E., Industrial Marketing Research, New York: John Wiley and Sons, 1979, pp. 374-415.

[6]Granbois, D., "Decision Processes for Major Durable Goods," in G. Fisk, ed., New Essays in Marketing Theory, Boston: Allyn and Bacon, 1971.

[7]Foote, Nelson N., "The Time Dimension and Consumer Behavior," in Joseph W. Newman (ed.), On Knowing the Consumer, New York: John Wiley & Sons, 1966.

[8]Synergic Resources Corporation (SRC), 1985a, Results of Kentucky Utilities Homebuilders Focus Group, Prepared for Kentucky Utilities, October 1985.

[9]Coughlin, R., "Understanding Fuel and Equipment Choice Decisions," Proceedings of the Great PGandE Expo, May 1985.

[10]Synergic Resources Corporation (SRC), 1985b, Results of Focus Groups With Architects, Engineers, Developers, Managers and Owners, Report prepared for North Carolina Alternative Energy Corporation, October 1985.

[11]McCutcheon, L. and C. McDonald, "Conservation Decisions in the Lumber and Wood Products Industry," Working Paper, submitted to Bonneville Power Administration, May 1986.

[12]Synergic Resources Corporation (SRC), 1985c, Marketing to the Commercial Sector, Seminar Manual, American Public Power Association, 1985.

[13]Synergic Resources Corporation (SRC), 1985d, Results of Focus Groups with Building Owners/Managers and Building Specifiers, Report prepared for Florida Power and Light Company, February 1985.

[14]Davis, Harry L., "Measurement of Husband-Wife Influence in Consumer Purchase Decisions," Journal of Marketing Research, 8, August 1971, pp. 305-312.

[15]Blood, Robert O. Jr. and Donald M. Wolfe, Husbands and Wives: The Dynamics of Married Living, Glencoe, Illinois: The Free Press, 1960.

[16]Sharp, Harry and Paul Mott, "Consumer Decisions in the Metropolitan Family," Journal of Marketing, 21, October 1956, pp. 149-156.

[17]Wolgast, Elizabeth H., "Do Husbands or Wives Make the Purchasing Decisions?," Journal of Marketing, 23, October 1958, pp. 151-181.

[18]Granbois, D. and Ronald P. Willett, "Equivalence of Family Role Measures Based on Husband and Wife Data," Journal of Marriage and the Family, 32, February 1970, pp. 68-72.

[19]Wilkening, E. A. and Denton E. Morrison, "A Comparison of Husband and Wife Responses Concerning Who Makes Farm and Home Decisions," Journal of Marriage and Family Living, 25, August 1963, pp. 349-351.

[20]Davis, Harry L., "Dimensions of Marital Roles in Consumer Decision Making," Journal of Marketing Research, 7, May 1970, pp. 168-177.

[21]McCann, Glen D., "Consumer Decisions in the Rural Family in the South," Paper presented at the Annual Meeting of the American Sociological Association, 1960.

[22]Scanzoni, John, "A Note on the Sufficiency of Wife Responses in Family Research," Pacific Sociological Review, 8, Fall 1965, pp. 109-115.

[23]Limaye, Dilip R., L. Andrews and C. McDonald, The Multiple Decision-Maker Problem, Report prepared for the Electric Power Research Institute, May 1986.

[24]Pollay, Richard W., "A Model of Family Decision Making," Working Paper No. 1, Lawrence: The University of Kansas School of Business, n.d.

[25]Ferguson, W., "A Critical Review of Recent Organizational Buying Research," Industrial Marketing Management, Vol. 7, pp. 225-230, 1978.

[26]Nicosia, F. and Y. Wind, "Emerging Models of Organizational Buying Processes," Industrial Marketing Management, Vol. 6, pp. 353-369, 1977.

[27]Wind, Y. and R. Thomas, "Conceptual and Methodological Issues in Organizational Buying Behavior," European Journal of Marketing, Vol. 14, pp. 239-263, 1980.

[28]Webster, F. and Y. Wind, Organizational Buying Behavior, Englewood Cliffs, New Jersey: Prentice-Hall, 1972.

[29]Kotler, P., Marketing for Nonprofit Organizations, Englewood Cliffs, New Jersey: Prentice-Hall, 1982, pp. 252-259.

[30]Robinson, Patrick J. and Charles W. Farris, Industrial Buying and Creative Marketing, Boston: Allyn and Bacon, 1976.

[31]Choffray, J. M. and G. Lilien, "Assessing Response to Industrial Marketing Strategy," Journal of Marketing, 42, April 1978, pp. 20-31.

[32]Sheth, J., "A Model of Industrial Buyer Behavior," Journal of Marketing, 37, October 1973, pp. 50-56.

[33]Wind, Y., "A Reward-Balance Model of Buying Behavior in Organizations," in G. Fisk, ed., New Essays in Marketing Theory, Boston: Allyn and Bacon, 1971, American Marketing Association, Chicago, 1978.

[34]Morris, M. and S. Freedman, "Coalitions in Organization Buying," Industrial Marketing Management, 13, 1984, pp. 123-132.

[35]Krapfel, R., "An Extended Interpersonal Influence Model of Organizational Behavior," Journal of Business Research, 10, 1982, pp. 147-157.

[36]Woodside, A. and P. Reingen, "Buyer-Seller Interactions: An Introduction" in Reingen, P. and A. Woodside, eds., Buyer-Seller Interactions: Empirical Research and Normative Issues, Chicago: American Marketing Association, 1981, pp. 1-10.

[37]Wilson, D., "Dyadic Interactions: Some Conceptualizations" in T. Bonoma and G. Zaltman, eds., Organizational Buying Behavior, Chicago: American Marketing Association, 1978.

[38]Bonoma, T. and W. Johnston, "The Social Psychology of Industrial Buying and Selling," Industrial Marketing Management, 17, 1978, pp. 213-224.

[39]Choffray, J. M. and G. Lilien, Market Planning For New Industrial Products, New York: John Wiley and Sons, 1980, pp. 135-154.

[40]Johnston, W. and R. Spekman, "Special Section on Industrial Buying Behavior: Introduction," Journal of Business Research, 10, 1982, pp. 133-134.

[41]Bellizzi, J. "Organizational Size and Buying Influences," Industrial Marketing Management, 10, 1981, pp. 17-21.

[42]Berkowitz, M., "New Product Adoption by the Buying Organization: Who are the Real Influencers?," Industrial Marketing Management, 15, 1986, pp. 33-43.

[43]Johnston, W., "Dyadic Communication Patterns in Industrial Buying Behavior," in Reingen, P. and A. Woodside, eds., Buyer-Seller Interactions: Empirical Research and Normative Issues, Chicago: American Marketing Association, 1981, pp. 11-22.

[44]Johnston, W., and T. Bonoma, "The Buying Center: Structure and Interaction Patterns," Journal of Marketing, 45, Summer 1981a, pp. 143-156.

[45]Johnston, W., "Purchase Process for Capital Equipment and Services," Industrial Marketing Management, 10, 1981b, pp. 253-264.

[46]Spekman, R. and L. Stern, "Environmental Uncertainty and Buying Group Structure: An Empirical Investigation," Journal of Marketing, 43, Spring 1979, pp. 54-64.

[47]Silk, Alvin J. and Manchar U. Kalwani, "Measuring Influence in Organizational Purchase Decisions," Journal of Marketing Research, May 1982, pp. 163-181.

[48]Munsinger, Gary M., Jean E. Weber and Richard W. Hansen, "Joint Home Purchasing Decisions by Husbands and Wives," Journal of Consumer Research, March 1975, pp. 60-66.

[49]Dwyer, F. Robert and Ann M. Welsh, "Environmental Relationships of the Internal Political Economy of Marketing Channels," Journal of Marketing Research, November 1985, pp. 397-414.

[50]Keeney, Ralph L. and Craig W. Kirkwood, "Group Decision-Making Using Cardinal Social-Welfare Functions," Management Science, December 1975, pp. 430-437. The specific functional models whose descriptions follow are not from published literature, but rather from proprietary studies conducted for major manufacturers.

[51]Anderson, James C., "A Measurement Model to Assess Measure-Specific Factors in Multiple-Informant Research," Journal of Marketing Research, February 1985, pp. 86-99.

[52]Lord, Charles G., "Schemes and Images as Memory Aids: Two Models of Processing Social Information," Journal of Personality and Social Psychology, V. 38(2), 1980, pp. 257-269.

[53]Rogers, T. B., N. A. Kuiper and W. S. Kirker, "Self-Reference and the Encoding of Personal Information," Journal of Personality and Social Psychology, V. 35(9), 1977, pp. 677-688.

[54]Phillips, Lynn W., "Assessing Measurement Error in Key Informant Reports: A Methodological Note on Organizational Analysis in Marketing," Journal of Marketing Research, November 1981, pp. 395-415.

[55]Krishnamurthi, Lakshman, "The Salience of Relevant Others and Its Effect on Individual and Joint Preferences: An Experimental Investigation," Journal of Consumer Research, June 1983, pp. 62-72.

[56]Synergic Resources Corporation (SRC), 1985e, Analysis of Commercial Sector Energy Decisions: Space Conditioning Fuel and Equipment, Final Report, prepared for New England Power Pool, October 1985.

[57]Synergic Resources Corporation (SRC), 1986a, Architect/Engineer Influence on Energy Systems Choice in New Commercial Building Design, Report prepared for Jersey Central Power and Light Company, February 1986.

[58]Synergic Resources Corporation (SRC), 1986b, Surveys and Interviews of Commercial/Industrial Decision-makers, Report prepared for Savannah Electric and Power Company, June 1985.

[59]Awad, Ziyad, "Industrial Market Research: New Directions for Customer Service," in Limaye, Sharon (ed.) Proceedings of the Electric Utility Market Research Symposium, Kansas City, MO, January 1987.

[60]Camera, Ken, and Dilip R. Limaye, Market Study of Thermal Energy Storage and Gas Air Conditioning, Report prepared for San Diego Gas and Electric Company, January 1986.

[61]Davis, Todd D., Development of a Strategic Marketing Plan for El Paso Electric Company, Report prepared for El Paso Electric Company, April 1984.

[62]Robinson Associates, Inc. (RAI), Study of Residential Appliance Purchase Decisions, Report prepared for Southern California Edison Company, 1985.

CHAPTER 6

Segmenting Commercial and Industrial Markets

Dilip R. Limaye

INTRODUCTION

As electric utilities confront the dual challenges of meeting customer demand and controlling costs, they are turning increasingly toward demand-side resources. Successful demand-side management (DSM) strategies depend heavily on substantial customer acceptance levels. Consequently, minimizing the risk of market acceptance becomes a vital component of any demand-side program, as well as of reliable load forecasting, effective customer service, and strategic marketing campaigns. All of these activities require, a priori, a thorough and accurate understanding of customer characteristics, customer decision processes and criteria, and load patterns. Utilities have traditionally focused their demand-side efforts on the residential sector. However, since much of the potential for residential load modification has been tapped, planners are now addressing the commercial and industrial markets.

Utility managers who seek to understand their commercial/industrial customers face some formidable obstacles, including an extraordinary diversity of customer characteristics and load patterns, a lack of adequate data on end uses and load shapes, the complexity of decision processes, and difficulties in predicting customer response to utility programs.[1]

The most effective approach to these issues and to demand-side planning is to disaggregate or segment commercial and industrial markets in order to focus particular programs where they will be most acceptable (or meet the least resistance). This chapter discusses some utility-tested methods for segmenting the C/I market, explains how they are used in actual case studies, and concludes with three examples of market segmentation; one aimed at analyzing cool storage penetration, one geared toward strategic conservation potential, and the third focused on understanding industrial energy decisions.

CHARACTERISTICS OF COMMERCIAL AND INDUSTRIAL MARKETS

Commercial and industrial (C/I) customers account for over 60% of electricity sales in the U.S.,[2] and are usually much larger than residential customers (Figure 6-1). Therefore, on a per customer basis, the C/I market offers significantly greater potential for load modification. But C/I customers exhibit tremendous diversity in terms of their size, physical characteristics, operating profiles, end uses, fuel mix, energy intensities, load shapes, and decision processes and criteria.

The diversity of energy utilization is illustrated in Tables 6-1 and 6-2. Table 6-1 shows energy consumption per square foot for different types of commercial buildings,[3] which ranges from 7.0 to 29.3 kWh/sq. ft. for electricity, and 50,000 to 129,000 Btu/sq. ft. for gas. Table 6-2 shows energy use as a percentage of total production costs for different industry groups.[4] Electricity ranges from 0.4% to 4.1% of total production costs, while total energy consumption ranges from 1.3 to 11.8%. Moreover, even within a building type or industry group, there is significant variation in customer characteristics and energy consumption.

Also important for the utility planners and marketers is the fact that energy use patterns are significantly different among different customer types. For example, large office buildings have large cooling loads during daytime periods, hospitals have large loads throughout the day and night, and restaurants are likely to have peak loads during evening hours. Further, the applicability of certain demand-side technologies or programs, and the criteria used by decision-makers to accept these technologies or programs, also may vary. The design of effective strategies and programs to modify customer loads therefore requires careful segmentation.

Figure 6-1. U.S. TOTAL ELECTRICITY SALES AND SALES PER CUSTOMER BY CLASS

TOTAL ELECTRICITY SALES IN THE U.S.

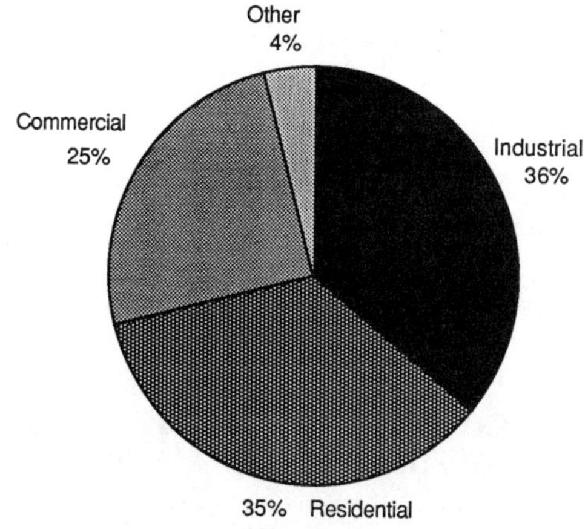

SALES PER CUSTOMER (MWH)

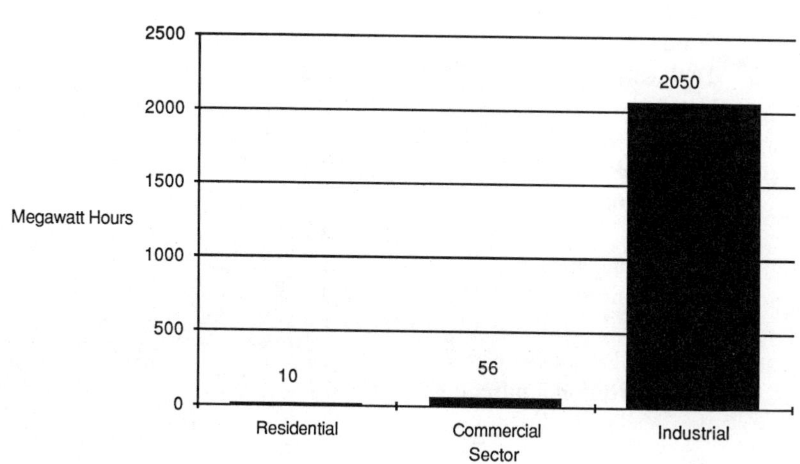

Table 6-1. ENERGY CONSUMPTION PER SQUARE FOOT FOR COMMERCIAL BUILDING TYPES

Building Type	Electricity (kWh)	Gas (Thousand Btu)
Food stores	29.3	119
Health care	20.2	125
Office	17.3	63
Hotel/motel	16.7	76
Warehouse	12.9	84
Retail	11.1	50
Education	8.2	52
Assembly	7.0	54
Automotive	10.3	97

Source: Energy Information Administration[3]

Table 6-2. ENERGY CONSUMPTION AS A PERCENT OF TOTAL PRODUCTION COST FOR DIFFERENT INDUSTRY GROUPS

Industry Group	Electricity %	Total Energy %
Food	0.9	2.0
Textiles	2.4	4.2
Lumber	1.5	3.0
Paper	3.0	9.1
Chemicals	3.8	8.5
Petroleum	0.7	2.3
Rubber	2.7	4.0
Stone, clay, glass	3.7	11.8
Primary metals	4.1	8.9
Fabricated metals	1.4	2.5
Electrical machinery	1.3	1.9
Transportation equipment	0.8	1.3

WHAT IS SEGMENTATION?

Segmentation involves disaggregation of a population into smaller groups and subgroups with some homogeneous characteristics. The purpose of segmentation is to be able to focus certain strategies and tactics on each group or subgroup, based on the characteristics of that group/subgroup, thereby achieving more effective results than if the strategies or tactics addressed the entire population.[5] Segmentation allows the planner/marketers to better:

- Understand and describe differences in market characteristics
- Analyze reasons for customer acceptance and response
- Predict the reactions of customers to utility programs/ incentives
- Influence or control acceptance and response
- Design more effective products, programs, and strategies.

Segmentation has been used very effectively by marketers in a number of different industries. Automobile manufacturers, for example, offer a wide variety of products with different types, styles, sizes, and performance characteristics, each appealing to a different market segment. Other consumer product manufacturers also offer a diverse range of products targeted at different consumer segments. Airlines have designed programs which target specific segments such as frequent fliers, tourists, senior citizens, etc.

Some electric utilities have segmented their residential markets for demand-side management. In one case, a northeastern utility targeted a heat pump program for suburban homeowners characterized by moderate to high incomes and oil-fired forced-air heating systems. Another utility in the Southwest designed a heat pump marketing program for affluent homeowners who placed a high value on cooling comfort. A southeastern utility successfully offered certain conservation programs to a segment comprising senior citizens. Similar segmentation and targeting can be developed for commercial and industrial markets.

METHODS FOR SEGMENTING COMMERCIAL AND INDUSTRIAL MARKETS

Because of the diverse characteristics of commercial and industrial customers, a number of different approaches to segmentation are possible, depending on the nature or purpose of the DSM or strategic marketing program.

Segmentation Based On SIC Codes

Perhaps the easiest method for segmenting commercial/industrial markets is by Standard Industrial Classification (SIC) codes. Most utilities have information on SIC codes in their billing records.

The advantage of this type of segmentation is that the SIC codes do classify relatively homogeneous customer groups. The problem is the very large number of segments. (There are 99 segments at the two-digit SIC level, and over a thousand at the four-digit level.) In addition, the SIC scheme is not designed from the perspective of energy consumption or load shapes, particularly for commercial customers.

Segmentation By Building or Industry Type

A second method of segmentation is by building or industry type. Commercial establishments can be segmented into major building groups (with several subgroups within some groups) as shown in Table 6-3. Similarly, the major industries can be grouped into a small number of segments, as shown in Table 6-4. SIC codes can be aggregated to develop these segments.

Table 6-3. LIST OF COMMERCIAL BUILDING TYPES

- **Office buildings**
  - small
  - large
- **Retail**
  - small
  - large
  - shopping center
- **Food stores**
  - small
  - supermarket
- **Schools**
  - elementary
  - secondary
  - higher education
- **Restaurants**
  - table service
  - fast-food
- **Health care**
  - hospital
  - nursing home
- **Hotels**
- **Motels**
- **Warehouses**
  - refrigerated
  - non-refrigerated
- **Civic centers and assembly buildings**
- **Movie theaters**
- **Churches**
- **Emergency services**
- **Automotive**
  - dealers
  - service stations
- **Miscellaneous**

Table 6-4. GROUPING OF INDUSTRY TYPES

Process Industries (Non-Metals Production)

20 Food and kindred products
21 Tobacco products
22 Textile mill products
26 Paper and allied products
28 Chemicals and allied products
29 Petroleum and coal products

Manufacturing Industries (Non-Metals Fabrication)

23 Apparel, textile products
24 Lumber and wood products
25 Furniture and fixtures
27 Printing and publishing
30 Rubber and miscellaneous plastics
31 Leather, leather products
32 Stone, clay, and glass products

Metals Production

33 Primary metals

Metals Fabrication

34 Fabricated metal products
35 Machinery
36 Electrical equipment
37 Transportation equipment
38 Instruments and related products
39 Miscellaneous manufacturing industries

Segmentation By Size

Further segmentation can be accomplished by using customer size. Size is often important because it is a determinant of customer characteristics and loads. For example, small office buildings have different physical characteristics and energy intensities than large offices. Similar differences exist between small and large retail establishments and food stores. (A key issue is defining the cut-off size for small versus large.) Size can also be an important determinant in the applicability of certain demand-side technologies (such as cogeneration and thermal storage), because of the economies of scale associated with such technologies. A measure of size can be obtained from billing data by using kW or kWh.

Segmentation By End-Use Load Patterns

Another approach to segmenting commercial and industrial markets involves studying the nature of certain end-use loads. For example, the best markets for cogeneration are customers with large, non-varying, low-pressure steam loads, with certain ratios of steam to electric loads. The best markets for thermal storage are customers with pronounced, short-duration cooling loads that are coincident with utility peak periods. Unfortunately, such load pattern data are not readily available.

Segmentation By Type of Decision-Maker

C/I markets may also be segmented according to the type of decision-maker who deals with energy-related issues. For example, in commercial buildings, decisions regarding investments in energy-using equipment may involve:

- Private owner/occupant
- Government owner/occupant
- Developers

- Architect/engineers
- Building managers
- Tenants.

This type of segmentation is useful in understanding the factors that influence energy investments.[6,7]

Segmentation Using Business or Industry Characteristics

Commercial/Industrial market segments can be defined using a combination of business/industry characteristics. Utility billing records will often contain some of this information. Some information can also be obtained from secondary sources such as Census data. However, it is usually necessary to conduct a customer survey in order to gather the needed information.

Typical characteristics that are useful in developing market segments include:

- SIC code
- Customer type and major energy end uses
- Energy consumption and energy demand
- Facility size
- Nature of certain end-use load patterns
- New versus old construction
- Region/location/climate
- Rate schedule

- Load shapes
- Electricity costs as a percentage of total operating expenses
- Ownership type
- Financial characteristics.

Psychographic Segmentation

While psychographics was originally used to describe residential customers, its application to business firms (sometimes called firmographics) is becoming widespread. Psychographics can be used to characterize organizations in terms of:

- Management structure and characteristics
- Complexity of energy decisions
- Bureaucracy
- Flexibility
- Entrepreneurism
- Purchase intentions
- Purchase behavior
- Motivations
- Awareness of and interest in certain products, programs, and features.

Attitudinal Segmentation

Attitudinal segmentation groups customers in terms of their attitudes toward the utility and toward specific types of programs/incentives. Measurable attitudinal variables include:

- Importance of energy in production process
- Levels and processes of decision-making
- Preferences for capital versus operating costs/savings
- Conservation attitudes
- Perceived importance of energy management
- Criteria used in purchase of equipment.

Examples Of Segmentation

Recent examples of C/I market segmentation include:

- Assessment of cool storage options for Florida Power & Light[8,9]
- Assessment of thermal storage, gas air conditioning, and cogeneration potential in commercial markets for San Diego Gas and Electric[10]
- Evaluation of commercial cogeneration for eight market segments for Consolidated Edison[11]
- Examination of cool storage in new and existing commercial buildings for the North Carolina Alternative Energy Corporation.[12]

Examples of attitudinal segmentation of C/I customers include recent research by Southern California Edison,[13] Duquesne Light Company,[14] and Jersey Central Power and Light Company.[14]

Two of these segmentation studies are briefly described below.

MARKET SEGMENTATION FOR COOL STORAGE ANALYSIS: CASE STUDY

Cool storage technology offers significant potential benefits to electric utilities and their customers. Using chilled-water or ice storage systems, this technology shifts peak period cooling loads to off-peak periods, thereby reducing utility production costs and offering customers reduced demand and (with time-of-use rates) energy charges. However, the application of cool storage technology is not equally beneficial to all C/I customers. In a recent study for Florida Power & Light (FPL), a market segmentation was performed to assess cool storage market potential.[8]

The first step in the segmentation process was an examination of the size distribution. An assessment of existing cool storage systems indicated that the technology was unlikely to be economical for customers with cooling loads smaller than 25 kW. Since data on customer cooling loads were not readily available, it was assumed that customers with total loads of 50 kW or less would be unattractive. The customers were therefore segmented by kW demand into two classes -- 50 kW or over, and under 50 kW.

The customers with loads of 50 kW or higher were segmented by building type or industry group. Fourteen groups of customers were identified as attractive markets for cool storage. Using SIC code information in the billing data file, the customers were segmented into these 14 groups (Table 6-5).

Table 6-5. MARKET SEGMENTS FOR COOL STORAGE ANALYSIS

No.	Type	SIC Codes
1.	Office buildings	40-49, 60-67, 73, 83, 91, 93-97
2.	Shopping centers	5318
3.	Other retail	52, 53 (excl. 5318), 55, 56, 57, 59, 72, 76
4.	Elementary/secondary schools	8211-8218
5.	Higher education	8221-8224, 824
6.	Hospitals	806
7.	Nursing homes/clinics	805, 807, 808, 809
8.	Hotels	7011
9.	Motels	7012
10.	Restaurants	58
11.	Food stores	54
12.	Warehouses	50, 51, 422
13.	Miscellaneous commercial	All others
14.	Assembly industries	34-38

Source: N. Bergstrom et al.[8]

To analyze the applicability of cool storage, a two-part survey sample of customers was conducted. The first part was a telephone survey of decision-makers to identify the major factors influencing decisions regarding HVAC equipment, the importance of each factor in the decision process, the methods and criteria used for economic analysis, information sources for new technology, and preferences for utility programs and incentives. A key finding of this survey was the distribution of desired payback, which was found to vary among different market segments (Figure 6-2). The second part of the survey addressed the physical, operating, and equipment characteristics of the building or plant, with emphasis on the cooling systems and equipment.

The survey results were analyzed and further segmentation was performed for office buildings (large versus small), retail (large versus small), and restaurant (fast food versus table service), yielding a total of 17 market segments. A prototypical building was then synthesized for each segment. Detailed hour-by-hour simulation of building loads (by end use) was then performed using a microcomputer-based model.

The applicability of cool storage was examined using another hour-by-hour simulation model of the cooling systems. The model is capable of addressing full, partial, or demand-limited storage. The simulation provided the size and operating characteristics of the optimal cool storage system, as well as its capital costs, for each prototype. A financial analysis was then performed to determine the changes in demand and energy charges, and the results were used in a market penetration model to determine the market potential and the time path of market penetration.

The market penetration analysis for each segment indicated that certain segments were far better candidates for cool storage than others.[9] The most attractive segments were:

- Large office buildings

- Large retail buildings

Figure 6-2. DISTRIBUTION OF DESIRED PAYBACK VARIES BY MARKET SEGMENT

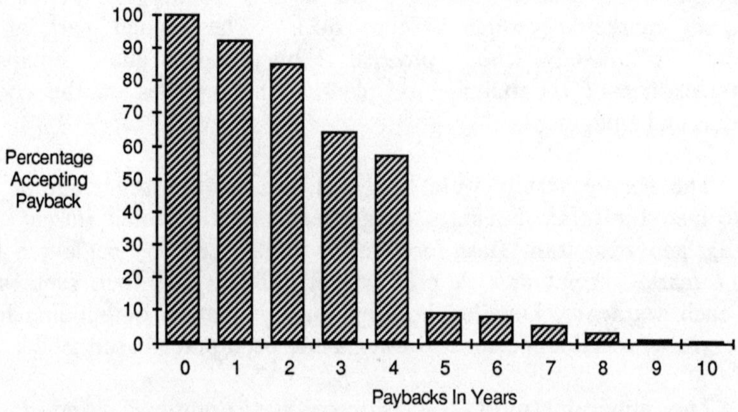

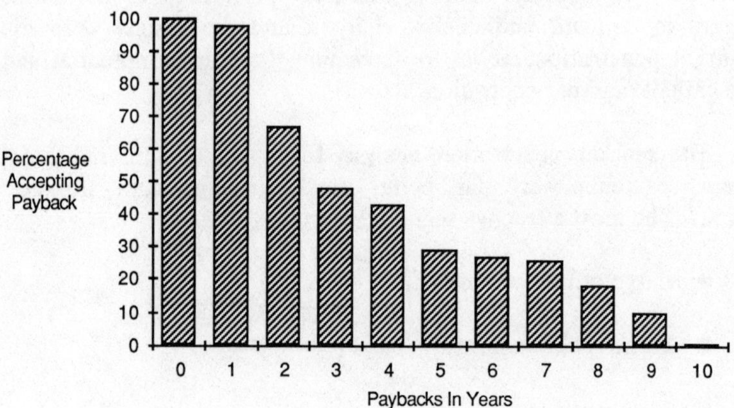

- Hospitals
- Schools
- Assembly industries.

In the final step of the analysis, different utility programs and incentives were examined and their incremental impacts on market penetration were calculated.

ATTITUDINAL AND BEHAVIORAL SEGMENTATION FOR STRATEGIC CONSERVATION: CASE STUDY

Jersey Central Power and Light (JCP&L) conducted a segmentation study for its commercial/industrial (C/I) market.[15] The objectives of this study were to

- Identify C/I market segments for strategic conservation
- Estimate the size of the segments
- Assess the key attitudinal and behavioral factors for each segment
- Determine the appropriateness of new conservation marketing strategies
- Identify and share "lessons of experience" in the conduct of C/I segmentation research.

Prior to this study, JCP&L has conducted a number of separate market research projects which evaluated reasons for acceptance/nonacceptance of specific conservation programs. A review of current conservation marketing activity indicated that:

- Most conservation programs were designed without systematic investigation of customer needs and benefits

- The current portfolio of conservation programs appeared to be competing for the same customers

- Some conservation programs may have outlived their useful lives and there may be a need to determine what program features would possibly bring about greater customer acceptance levels

- Better estimates of market size and customer acceptance were needed for budgeting and planning purposes

- There was a need to identify more reliably, in advance of program implementation, key attitudinal and behavioral business characteristics and associated marketing program options.

The segmentation was accomplished using a telephone survey of 600 customers. The survey covered such topics as business characteristics (e.g., employment size, type of business, type of facilities, etc.), end uses, energy decision processes and criteria, awareness/interest in current and possible new JCP&L programs, and attitudes toward the utility.

The BMDP applications software program was used to analyze the survey data, using the following statistical techniques:

- **Factor analysis (maximum likelihood)** -- to reduce the number of variables to a few more meaningful dimensions or factors and identify their key segment attributes

- **Cluster analysis** -- based on standardized factor scores to determine the similarity of customer subgroups and to classify customers where intergroup distances are large and intragroup distances are small

- **Discriminant analysis** -- to help identify target markets by providing a linkage between psychographic variables and more tangible variables such as demographics, firmographics, energy usage, etc.

The analysis of survey data indicated nine key attitudinal factors that could be used to define market segments. These factors are shown in Table 6-6. The discriminant analysis provided three attitudinal clusters -- "low potential," "blocked potential," and "already conserving" (Table 6-7).

The "low potential" segment is distinguished by its nonbureaucratic features and its moderate identification with "not being able to save money by conserving." Some innovation also characterizes this segment. This information suggests that while no internal barriers exist to prevent investing in energy conservation, the real concern is that no significant benefits result from conserving.

The "blocked potential" segment feels that they are wasting energy, that they are very bureaucratic, and that there are other barriers to conservation. Conversely, members of the "already conserving" segment do not feel that they are wasting energy, but they, too, feel bureaucratic. Surprisingly, the "already conserving" segment scores lower on innovation.

The statistical approach was replicated for a set of behavioral variables. Factor analysis was used to identify the more important factors and the amount of explained variance. Table 6-8 identifies eight behavioral factors. The first factor, "interest in participating in new programs," represents interest in four new program concepts that respondents were asked to evaluate.

"Industrial energy management conservation steps taken or planned," the second factor, represents a sizeable increment in the explained variance (+12%). The remaining six factors represent commercial conservation steps taken/planned, total program participation, and interest in four different existing programs (e.g., lighting ballast, audit, high efficiency A/C, and time-of-use rates).

Table 6-6. ATTITUDE FACTORS WHICH DEFINE SEGMENTS

Segmentation/ Factor	Key Variable Descriptors (Loading)
1. Economic payback	• Payback method (+.898) • Rate of return (-.781)
2. Non-bureaucratic corporate structure	• Few job titles (+.740) • Wear many hats (+.645) • No R&D (+.358)
3. No barriers to conservation	• Capital is not a barrier (+.577) • No red tape for new technology (+.482) • Priorities are well defined (+.413) • Price volatility no problem (+.404) • Can get good information (+.309)
4. Social setting in corporation	• Poor cooperation (+.613) • Unfriendly atmosphere (+.565)
5. Cannot save by conserving	• Better opportunities for saving money other than energy management (+.604) • Cost of energy management won't provide worthwhile savings (+.458) • Do monitor energy use (-.372)
6. Net present value	• Use net present value (+.847)
7. Have taken conservation steps	• All energy savings steps taken (+.775) • No part of business wasting energy (-.311)
8. Unclear evaluation criteria	• Don't know economic evaluation methods used (+.677)
9. Innovation	• High R&D investment (+.519) • Innovative firm (+.487)

Table 6-7. MAJOR FACTOR DESCRIPTORS OF ATTITUDINAL CLUSTERS (CLUSTER MEANS)*

Segment Descriptors	Cluster 1 "Low Potential"	Cluster 2 "Blocked Potential"	Cluster 3 "Already Conserving"	F Ratio**
Non-bureaucratic	.9962	-.4722	-.3503	440.893
No barriers to conservation	.0042	-.2841	.2280	24.423
Unfriendly/ uncooperative atmosphere	.0105	.1722	-.1479	9.340
Not saving by conserving	-.3600	.1913	.1098	32.436
Feel wasting energy	.0392	.7031	-.6009	261.563
Innovation	.2803	.0754	-.2680	38.303

*Mean factor score on cluster

**Differences among the clusters with respect to factors is significant

-.1 to .1	Neutral on factor
.1 to .3	Moderate on factor
> .3	Heavy on factor

Table 6-8. BEHAVIORAL SEGMENTS

Segment/Factor	Key Variable Descriptors
1. Interest in participating in new programs	• Energy management control systems -- $10,000 rebate (+.664) • 20% rebate of energy savings up to $5,000 (+.648) • Shared savings interest (+.603) • Rebate for thermal (+.572) or cooling system (+.540)
2. Industrial steps taken or planned	• Efficient motor steps taken/planned (+.617) • Proportion of end uses for which conservation steps were taken (+.525) • Water heating steps taken/planned (+.444) • Business machines (+.416) • Industrial furnaces and process heat (+.407) • Refrigeration steps taken/planned (+.382) • Boiler/process steam steps taken/planned (+.375) • Laundering steps taken/planned (+.318)
3. Lighting/ballast interest (existing programs)	• Interested in lighting program (+.881) • Interested in ballast program (+.828) • Number of existing programs of interest (+.606)
4. Commercial conservation steps taken or planned	• Space cooling steps taken/planned (+.647) • Space heating steps taken/planned (+.633) • Lighting steps taken/planned (+.468) • Water heating steps taken/planned (+.401) • Cooking steps taken/planned (+.346) • Refrigeration steps taken/planned (+.337)
5. Audit interest	• Interested in energy audits (+.912) • Number of existing programs of interest (+.413)
6. Program participation	• Participation level in existing programs (+.959)
7. High efficiency A/C program interest	• Interest in participating in A/C rebate program (+.898)
9. TOU interest	• Number of existing programs are of interest (+.373)

Cluster analysis was then used to identify four significant behavioral segments (Table 6-9). The "past participants" segment is characterized heavily by the level of total past participation in JCP&L programs and by participation and interest in the JCP&L lighting/ballast and other programs. This segment has also expressed interest in both commercial and industrial conservation steps. The segment labeled "interest in existing programs but not new programs" is characterized as having significant interest in the high efficiency A/C rebate program and in the lighting/ballast rebate program. This segment also has a low score on total program participation. The "interest in new programs" segment consists of customers who have not participated in prior programs, including energy audits, but who are interested in participating in some of the new program options. The "interest in existing lighting programs" segment is distinguished by particularly high interest in the lighting/ballast program. However, there is strong disinterest in the high efficiency A/C rebate program and a lower level of prior program participation.

Significant characteristics that separate the four segments are varying levels of consumption, varying decision-maker roles and responsibilities, and differing business activity. Similarly, end uses, energy cost impacts, and the number of facilities operated also vary.

A cross-tabulation of attitude segments versus behavioral segments is presented in Table 6-10. This table is important because it identifies the proportion of the JCP&L customer population in each segment. A total of 39% of customers feel that they are "already conserving." "Low potential" represents 29% of the customer population, and "blocked potential" represents 32%.

In terms of behavioral clusters, past program participants represent 10% of C/I customers. Those interested in new programs constitute 44% of C/I customers, and those interested in existing and new programs represent 19% and 26% of C/I customers, respectively.

Table 6-9. MAJOR FACTOR DESCRIPTORS OF BEHAVIORAL CLUSTERS (CLUSTER MEANS)*

Factors	Cluster 1 "Past Participants" (11%) (N=65)	Cluster 2 "Interest in Existing Programs but Not New Programs" (19%) (N=116)	Cluster 3 "Interest in New Programs" (44%) (N=157)	Cluster 4 "Interest in Existing Lighting Programs" (26%) (N=157)
Lighting/ballast interest -- existing programs	.5738	.7513	-.9716	.8411
Commercial conservation steps taken/planned -- existing programs	.3054	.0911	-.0804	-.0585
Interest in participating in new programs	.0706	-.2629	.1563	-.0344
Industrial steps taken/planned	.3185	.1211	-.1049	-.0540
Audit program interest	.1117	.3503	-.1327	-.0080
Participation in existing programs	2.3241	-.4733	-.1484	-.3629

Table 6-10. CROSS TABULATIONS OF ATTITUDE AND BEHAVIORAL CLUSTERS

	BEHAVIORAL CLUSTERS				
Attitude Clusters	Past Participants	Interest in Existing Programs but not New Programs	Interest in New Programs Only	Interest in Existing Lighting Programs	Total
Low potential	1% (9)	5% (31)	16% (95)	6% (39)	29% (174)
Blocked potential	4% (24)	9% (52)	11% (66)	8% (50)	32% (192)
Already conserving	5% (32)	5% (33)	17% (103)	11% (68)	39% (236)
Total	11% (65)	19% (116)	44% (264)	26% (157)	100% (602)

Clearly, not all of those expressing interest will in fact participate in JCP&L programs. This is why it is important to look at the underlying attitudinal dimensions of these behavioral segments. Table 6-11 identifies the possible programs that could be implemented, given the segments identified. The lighting/ballast program appears to be suitable for implementation across a wide range of market segments, including those interested in existing lighting programs, those interested only in existing programs, and past program participants. For those customers interested only in new programs, rebates for energy management control systems, shared savings, and 20% rebates up to $5,000 are also appropriate.

Table 6-11. POSSIBLE PROGRAMS FOR SELECTED C/I SEGMENTS

Behavioral Segment	Attitudinal Segment	Segment Size	Programs
Interest in existing lighting programs	Low potential; already conserving	17%	Lighting/ballast rebates
Interest only in new programs	Already conserving	18%	EMCS rebates, 20% rebate of energy savings up to $5k, shared savings, thermal storage rebates
Interest in existing programs only	Blocked potential	9%	High efficiency A/C program and lighting ballast rebates
Past participants	Blocked potential; already conserving	9%	Lighting/ballast rebates, industrial energy management, general commercial energy management, time-of-use rates

UNDERSTANDING INDUSTRIAL ENERGY DECISIONS: CASE STUDY

The Bonneville Power Administration (BPA) has embarked upon major efforts to promote conservation in the industrial sector. BPA's research has shown that the industrial sector offers a significant potential for modification of customer loads and energy consumption in a manner that will benefit both the customer and the region. The lumber and wood products industry is one of the five major industries in the Northwest with the greatest energy conservation potential. However, the key to being able to achieve this potential lies in understanding the decision-making criteria and processes that exist in this market.

BPA recently explored the decision processes and criteria used by wood and lumber businesses to make decisions related to energy investments.[16] The research addressed the following questions:

- How are decisions made for the application of conservation, load management, and other demand-side management options?

- How do the decision processes relate to ownership patterns?

- What can BPA do to shape the conservation marketing program for this industry?

The research tools used included two focus groups with decision-makers representing small and large firms and a telephone survey of 200 firms. Site-specific data on employment, gross sales, and energy consumption were also obtained from a smaller sample of firms.

The lumber and wood products industry was first segmented by the type of product, and six segments were identified for further study based on their relatively high energy use:

- Sawmills and planning mills, general (SIC 2421)

- Hardwood dimension and flooring mills (SIC 2426)

- Mill work (SIC 2431)

- Hardwood veneer and plywood (SIC 2435)
- Softwood veneer and plywood (SIC 2436)
- Particle board (SIC 2492).

A short telephone survey of approximately 25 questions was administered to a random sample of mills in the six segments. The main purpose of the survey was to determine ownership patterns, since it was assumed that type of ownership is a major factor affecting how decisions are made. Four different ownership types were identified:

- Wholly-owned corporation
- Subsidiary
- Partnership
- Family-owned business.

The implications of the ownership type in decision-making were further explored in the two focus groups.

The results of this research found that mills of all types and sizes have similar decision-making processes and criteria, largely because they are operating in the same competitive environment. Substantial differences exist, however, depending on whether decisions are being made in a corporate or noncorporate environment. The two environments, one more formal and the other less formal, have an influence on the degree of analysis performed and the speed with which decisions are made.

The first two ownership types (representing corporate environments) involve more complicated decision-making processes, since hierarchies of authority in these businesses tend to be more complex. They tend to have a complex corporate structure with formal systems for decision-making. Mill engineers, mill managers, financial executives, corporate staff, and corporate executives may be involved in various aspects of the decision-making process. The complexity of the decision-making process and the number of different decision-makers involved are likely to be greater for decisions involving larger amounts of capital and/or operating expenses.

In contrast, decision-making by one person most often takes place in small firms, generally partnerships or family-owned businesses, where decisions can be made more rapidly. Often, the owner is the chief supervisor, treasurer, mechanic, and president, or is at least in close contact with each. Knowledge of all phases of the firm's performance and options is concentrated in one or a few individuals, thereby reducing some of the complexities in decision-making.

The distribution of the mills by ownership type is shown in Figure 6-3. Information was also collected on the location of decision-making activities. The respondents were asked whether capital decisions were made on-site or elsewhere; and if elsewhere, whether they were made locally, in the same city, or at the regional or national headquarters. Information on how decision-making patterns are distributed by ownership type is provided in Figures 6-4 and 6-5. Within each ownership type, at least 50% of the respondents reported that capital investment decisions were made on-site. In fact, decisions in over 70% of nonsubsidiary corporate mills were reportedly made on-site. Subsidiaries tend to have the most centralized decision-making procedures, reflected by the result that in half of these mills decisions are made somewhere other than at the site.

Figure 6-3. DISTRIBUTION OF MILLS BY OWNERSHIP TYPE

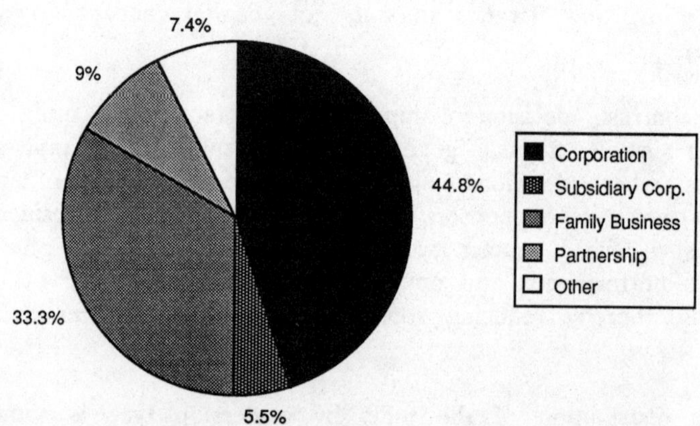

Figure 6-4. WHERE MAJOR CAPITAL DECISIONS ARE MADE BY OWNERSHIP TYPE

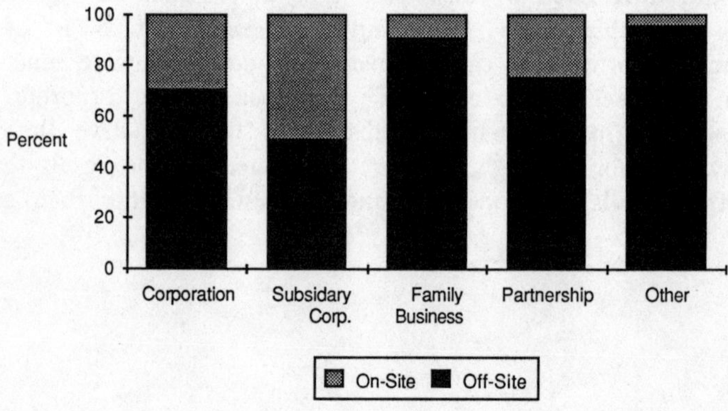

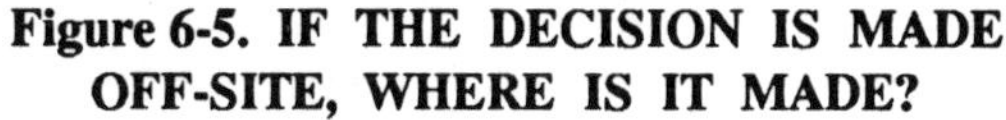
Figure 6-5. IF THE DECISION IS MADE OFF-SITE, WHERE IS IT MADE?

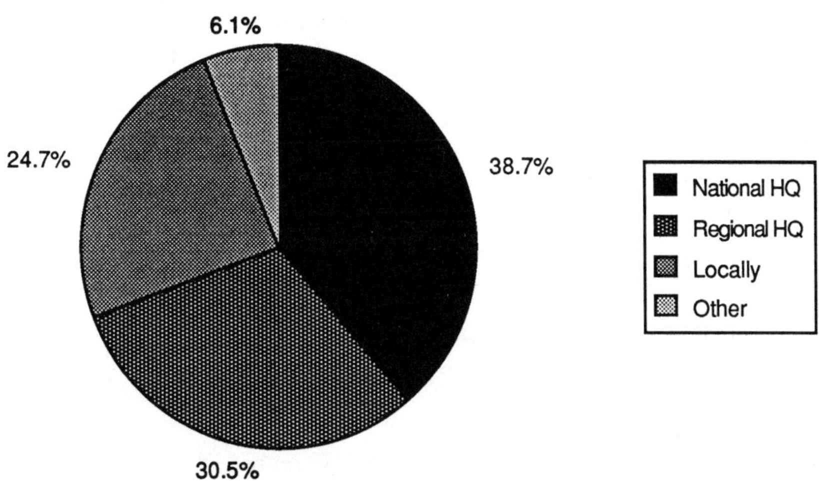

If decisions are made elsewhere, they tend to be made at least regionally or locally (31 and 25% of the decisions made elsewhere, respectively). However, 39% of the mills making decisions elsewhere make capital decisions at the national headquarters. Table 6-12 shows additional information on the decision-making process.

The results of the segmentation by ownership type lead to the following implications for BPA's conservation programs:

- Approximately 76% of all decisions are made at the plant level, indicating the importance of focusing on the plant manager or engineer in promoting conservation.

- It is easier to reach the decision-maker in the family-owned or partnership type of businesses than the corporate-owned or subsidiary firms.

Table 6-12. WHERE CAPITAL DECISIONS ARE MADE BY OWNERSHIP TYPE

Ownership Type	Number	At Plant	Locally	Regional Headquarters	National Headquarters	Elsewhere
Wholly-owned corporation	82	74.4%	4.9%	8.5%	9.8%	2.4%
Subsidiary corporation	10	50.0%	0.0%	10.0%	40.0%	0.0%
Family-owned	65	83.1%	9.2%	3.1%	4.6%	0.0%
Partnership	18	72.2%	5.6%	16.7%	0.0%	5.6%
Other	14	85.7%	0.0%	7.1%	7.1%	0.0%
All Types	189	76.7%	5.8%	7.4%	8.5%	1.6%

- Conservation marketing programs should be targeted at the plant supervisors, owners, or managers of family-owned or partnership type firms.

- For corporate-owned firms, BPA must target not only the local plant managers or conservation staff but also the regional or national headquarters staff that may influence energy decisions.

- Subsidiary corporations are likely to be the most difficult to reach for promoting conservation, but they represent a small proportion (5.5%) of all mills. It would therefore not be desirable to develop customized programs for these types of firms.

- In implementing specific programs, BPA should try to develop a better information base on ownership and decision processes for individual firms.

CONCLUDING REMARKS

Market segmentation can be an effective and valuable technique for analyzing and targeting demand-side management options for commercial and industrial customers. A wide range of segmentation methods and techniques can be employed, depending on the specific study needs and the characteristics of the customers served. While a detailed market segmentation study will generally require significant data collection, research, and analysis efforts, it will also allow more precise targeting of programs, resulting in greater customer participation. Indeed, the utility benefits of segmentation are likely to far exceed the costs.

REFERENCES

[1]Battelle-Columbus Division and Synergic Resources Corporation, Demand-Side Management: Volume 4, prepared for Electric Power Research Institute and Edison Electric Institute, 1986.

[2]Energy Information Administration, Electric Power Annual, Washington, D.C., 1985.

[3]Energy Information Administration, Non-Residential Building Energy Consumption Survey, EIA-0318-1.

[4]Synergic Resources Corporation, "Electricity Use and Production Costs in Manufacturing," Working Paper, prepared for Electric Power Research Institute, 1986.

[5]Feldman, S. "Segmentation," in Proceedings of Electric Utility Market Research Symposium, EPRI EA-4338, 1986.

[6]Limaye, D.R., "Assessing Market Penetration of New Technologies in the Commercial Sector," paper presented at the PGandE Energy Expo '85.

[7]McDonald, C. and L. Andrews, "Overview of the Commercial Investment Decision Process and Models," paper presented at IEEE Summer Conference, Vancouver B.C., June 1985.

[8]Bergstrom, N, et al, Cool Storage Market Analysis: Technical Discussion of Analysis Approach, prepared for Florida Power & Light, 1985.

[9]Synergic Resources Corporation, Cool Storage Market Potential in FPL Service Area, Draft Final Report, prepared for Florida Power & Light Company, November 1985.

[10]Limaye, D.R. and R. Crane, Assessment of Thermal Energy Storage and Gas Air Conditioning, report prepared for San Diego Gas and Electric Company, January 1985.

[11]Synergic Resources Corporation, Cogeneration Potential Assessment Model, Final Report, prepared for Consolidated Edison, July 1985.

[12]Synergic Resources Corporation, Assessment of Cool Storage in Commercial and Institutional Buildings, Phase I Report, prepared for North Carolina Alternative Energy Corporation, January 1986.

[13]William, M.V., "Segmenting Markets for Conservation," in Proceedings of the Conference on Utility Conservation Programs, EPRI EA-3530, May 1984.

[14]Feldman, S., "Determinants of Interest in New Energy Reduction Projects Among Commercial and Industrial Ratepayers," in Proceedings of the Conference on Utility Conservation Programs, EPRI EA-3530, May 1984.

[15]Lloyd, Gayle and Todd D. Davis, "Identifying Commercial/Industrial Market Segments for Utility Demand-Side Programs", Paper presented at the Great PGandE Energy Expo, Oakland CA, 1986.

[16]Synergic Resources Corporation, Conservation Decision-Making in the Lumber and Wood Products Industry, Final Report, submitted to Bonneville Power Administration, Report 7283-R1, November 1983.

CHAPTER 7

Test Marketing Demand-Side Programs

Todd D. Davis and Carol Sabo

WHY TEST MARKET?

The electric utility industry is currently undergoing a fundamental transition. As a growing, capital-intensive industry, utilities historically placed heavy emphasis on the production side of the business. As a result, over the years the electric utility industry was able to gain significant experience with the supply side of the business. The risks associated with actions on the supply side were generally known, and thanks to the high growth that was occurring in the boom years of the industry, precise evaluation of the performance of marketing programs was not that critical. Moreover, the relative value added of marketing programs was limited compared to the growth contributed by the expanding economy.

Now it could be argued that the electric utility industry is moving into a new frontier. For some electric utilities the greatest value added for electric service may come from the demand side of the business. That is, whether a utility's load shape objective is strategic load growth or conservation, future customer benefits may result largely from how well a utility is able to market to its customers. This turnabout stems largely from the fact that the electric utility industry has matured. Uncontrolled growth in the future may lead to poor financial performance and perhaps higher rates.

It could also be said that there is much risk on the demand side of the business. Just as power plant superintendents are concerned with the efficient operation of power plants, so should marketing managers be concerned about the efficiency and effectiveness of marketing programs -- efficiency in terms of using the lowest level of resources required to achieve a marketing objective, effectiveness in terms of the achievement of planned objectives. Test marketing can be helpful in that programs are designed on a smaller scale in order to judge how well they perform and to provide an opportunity to make the necessary adjustments before the program is implemented on a full-scale basis. In the consumer products industry, it is generally believed that nine out of ten new products fail. This failure may be due either to the product entering the market too soon, or to the use of the wrong marketing mix, or to the wrong market being targeted. If electric utility marketing programs are to be relied upon to manage the demand that customers place on the system, and if long-range planning decisions regarding capacity requirements are to be made based on the effects of utility marketing programs, then some of the same standards of reliability and the corresponding levels of monitoring that are standard in the consumer products industry should be applied to utility marketing programs. It is only through test marketing that a utility can determine the most likely performance of a program, because it provides an opportunity for pretesting marketing programs in a setting that comes the closest to approximating real market conditions.

WHAT IS TEST MARKETING?

Test marketing represents the systematic evaluation of utility marketing programs in a setting that comes close to approximating relevant conditions in which a product, market implementation methods, and customers interact. There are two primary functions that test marketing performs:

- Forecasting market acceptance of a product and other elements of the marketing strategy

- Evaluating the relative performance of the market implementation techniques.

Conceivably, test marketing can apply to every aspect of utility marketing. The level of resources devoted to test marketing, however, should vary, and should be proportional to the total risk associated with the marketing effort or the total value of expected benefits from the marketing effort.

Electric utilities are likely to encounter a number of unique problems in developing test marketing programs. Electric utilities will find, as many consumer durable manufacturers have, that scaling up from a test marketing program may not be that major a leap considering the level of resources needed to run a test marketing program. However, as will be argued in the final section of this chapter, there is still a need to seriously consider test marketing as a key part of marketing program planning. Of course, it must be remembered that the degree of influence utilities will have on product design and distribution may be minimal. This problem may threaten the validity of a test. Utility planners may find that a good proportion of time will be devoted to managing problems that arise to threaten the validity of a market test.

There are a number of potential marketing situations that an electric utility may want to test. Table 7-1 identifies a selected number of those activities which would serve as good candidates for test marketing. Advertising is an ideal marketing activity that utilities can and should test. There is much uncertainty regarding how effective advertising is in stimulating customers to act, yet for some utility marketing programs, advertising represents the largest expenditure element. Typically, an advertising agency is relied upon to test the concept and communication materials. Mall intercepts are generally used to test the communication. Sometimes utilities will use reach and frequency surveys to track how well an ad is doing. This form of testing may be sufficient if advertising is the only form of marketing that is used to support a program.

Table 7-1. GENERIC MARKETING CONCEPTS

1. DESIRED PRODUCT POSITIONING STRATEGIES

- Technical performance
  - heat only
  - dual fuel
  - heating/cooling
  - high efficiency
- First cost vs. life cycle cost
- Desired energy form
- Importance of brand
- Reliability/risk
- Safety
- Color
- Size

2. MARKET OCCASIONS

- Time-of-use
- New vs. replacement market
- Seasonality
- Pricing periods
- Primary vs. supplemental

3. MARKETING STRATEGIES

- Communication (i.e., type of media, copy testing, etc.)
- Promotion (i.e., rate design, incentive level and type, etc.)
- Trade allies (i.e., size, type, and relative effectiveness)
- Control strategies (i.e., short-, medium-, and long-duration) and customer acceptance
- Direct contact (i.e., type, location, etc.)

There may be occasions when a utility must balance the cost of an intensive advertising campaign versus the use of financial incentives. A utility may also want to evaluate the relative effectiveness of different levels of financial incentives (i.e., low, medium, and high rebates). There may also be a circumstance where a utility wants to evaluate the interactive effects of a rebate and paid advertising. This article presents a framework and hypothetical analysis that a utility may follow to develop a test marketing program that allows for the testing of two or more of the marketing methods being used.

WHAT IS AN APPROPRIATE TEST MARKETING FRAMEWORK TO APPLY?

A general framework that utilities may follow in approaching the design of a test marketing program is presented in Figure 7-1. The recommended approach involves six steps. For illustrative purposes examples will be used for a utility that recently completed a test marketing program for high efficiency appliances. A short description of each step in developing a test marketing program is provided.

Specification of Program Objectives

The specification of program objectives is one of the most important steps in the process. Program objectives should be specified clearly in order to provide guidance in the formulation of the major research questions. An example of an appropriate program objective is the following:

> The purpose of this project is to identify the levels of customer participation in a high efficiency appliance marketing program resulting from the use of varying levels of financial incentives versus the use of an advertising-only marketing strategy. In addition, this project is to identify the interactive effects of advertising and financial incentives in terms of gaining market acceptance. Finally, the relative role and involvement of trade allies in gaining customer awareness and acceptance of the desired products need to be identified.

Figure 7-1. TEST MARKETING FRAMEWORK FOR UTILITY DEMAND-SIDE PROGRAMS

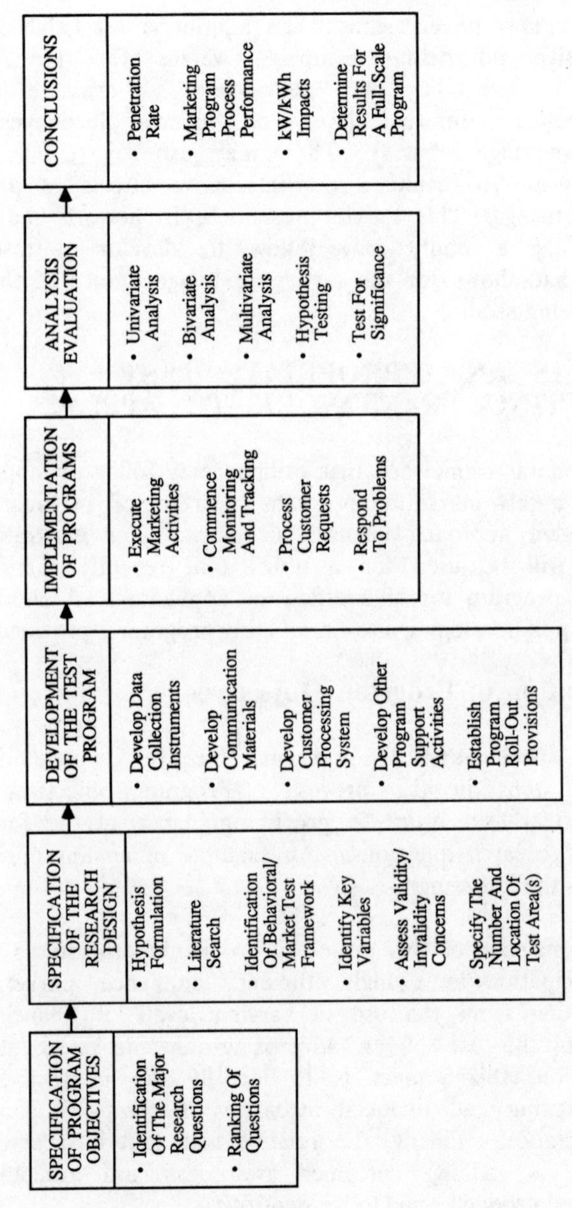

If a utility were interested in testing the marketing and customer acceptance of a direct load control program, the following objectives could be stated:

> The purpose of this project is to identify the relative levels of participation resulting from low and high monthly bill credits, coupled with two types of communication strategies. The first type of communication strategy pertains to using personalized direct mail and the second communication strategy consists of the use of doorknob hangers and selected door-to-door canvassing.

Once the statement of objectives has been formulated, a series of research questions can be generated that may eventually be restated as research hypotheses. The specific research questions that may be asked for the appliance rebate and communication program are as follows:

- Do financial incentives of low and high levels result in corresponding low and high levels of customer participation?
- Do financial incentives have a greater impact on increasing customer acceptance than a communication program?
- Does the influence of trade allies vary between incentive and communication programs?
- Are alternative types of incentives and communication programs received differently by customers having distinct demographic characteristics?
- Do trade allies behave differently in communication programs than they do in rebate and financial incentive programs?

The number of research questions that could be raised may be substantial. Thus, an immediate task for the utility planner is to first raise the questions, then rank the questions in their order of importance. When the ranking of these questions is completed, the utility planner is ready to specify the major hypotheses to be tested.

Specification of the Research Design

In specifying the appropriate research design, the utility planner must address the major research hypotheses or, to put it another way, determine the relationships among the variables being tested. To help identify the possible relationships that may exist in any new research project, one could:

- Contact other utilities that have implemented similar programs and identify their experiences.
- Review the relevant literature on utility experiences with similar programs and identify key variables and findings.
- Conduct qualitative research involving the target market in order to identify the key acceptance factors associated with the product and preferences related to alternative marketing strategies.

The hypotheses being tested should clearly state the possible relationships among the variables being tested. The following are examples of the hypotheses that may be tested in the hypothetical rebate and communication program:

- Communication-only programs will result in a lower customer acceptance of the technology being promoted than programs that use financial incentives.
- Trade ally acceptance of the marketing program will vary by the marketing mix that is employed.
- The higher the rebate (i.e., \$0, \$35, or \$50), the higher the participation level.

- The effectiveness of point-of-purchase materials will be dependent upon the degree of dealer support of the program.

- There will be a significant variation in how certain demographic groupings of customers will respond to the program.

The research hypotheses presented above now provide an opportunity to determine the key variables that should be tested, and the specifications of the major target population groups. Given the hypotheses presented above, the major population groups are utility customers and trade allies.

Next, there should be a definition of some of the factors that may intervene and possibly bias the results of the study. A number of factors could influence the results of our hypothetical study:

- Significant changes in customer income and employment that affect technology purchase patterns may occur.

- The planned market test period may be influenced by seasonal sales patterns.

- The current stock of appliances that dealers have on hand (rather than newer high efficiency models) may be the first offered to customers, thus "masking" the effects of the marketing program.

- Climatic factors may have a significant effect on customer purchase patterns.

Each of these intervening factors should be evaluated in terms of its potential threat to the validity of the study results. There are two types of validity that, if not present, can jeopardize the results of a test marketing study:

- **Internal validity** -- the ability to compare across treatment and control areas

- **External validity** -- the ability to generalize from the test population to the full population.

A number of factors may cause questions regarding internal validity to occur. These include self-selection bias, changes in the before versus after conditions in the market, and test bias. Problems with external validity are caused to a large extent by test market populations that are not representative of the larger market for which the full-scale program will be implemented. Utilities must be careful when selecting test markets to be sure that they do represent conditions similar to those of the market to which the full-scale program will be directed.

Table 7-2 presents a listing of both independent and dependent variables to be measured in our hypothetical study. Independent variables refer to the factors that usually precede the outcome measures in time. For example, customer demographics, attitudes, and marketing strategies precede customer participation in time. The outcome measures are usually the dependent variables; i.e., the participation of a customer in a utility's program is dependent upon demographics, attitudes, and the presence or absence of marketing variables. An illustrative framework is presented in Figure 7-2 showing the relationship of independent and dependent variables in a test marketing experiment.

The number of test areas to be used is also an important consideration. There may be an occasion when both the test and experimental groups are located in the same geographic area. A simulated market test allows a utility an opportunity to alter the test stimuli for randomly selected customers and then record the reported intentions of customers. Direct mail and direct contact represent the major forms of customer contact in conducting such studies. Figure 7-3 represents an example of how a simulated market test may be applied to a commercial marketing program. The number of customers included in such a test depends on the desired precision level and the variance of the data collected.

Table 7-2. EXPERIMENTAL UNITS AND INDEPENDENT/DEPENDENT VARIABLES IN UTILITY TEST MARKETING PROJECT

	Test Variables	
Experimental Units	Independent	Dependent
Customers	Demographics Attitudes Marketing mix	Recall/Awareness Characteristics of units sold Penetration rate Preference changes
Stores	Market size Store type Marketing mix	Sales
Sales Territories	Customer characteristics Marketing mix	Participation Consumption Load profile

The major advantages of a simulated market test are the following:

- Estimates of program participation under full knowledge and awareness are obtained.
- More concepts may be tested than is feasible in a larger field test.
- Estimates of price/incentive elasticity may be tested using split-half samples.

Figure 7-2
BASIC TEST MARKETING PROBLEM AND RESEARCH DESIGN

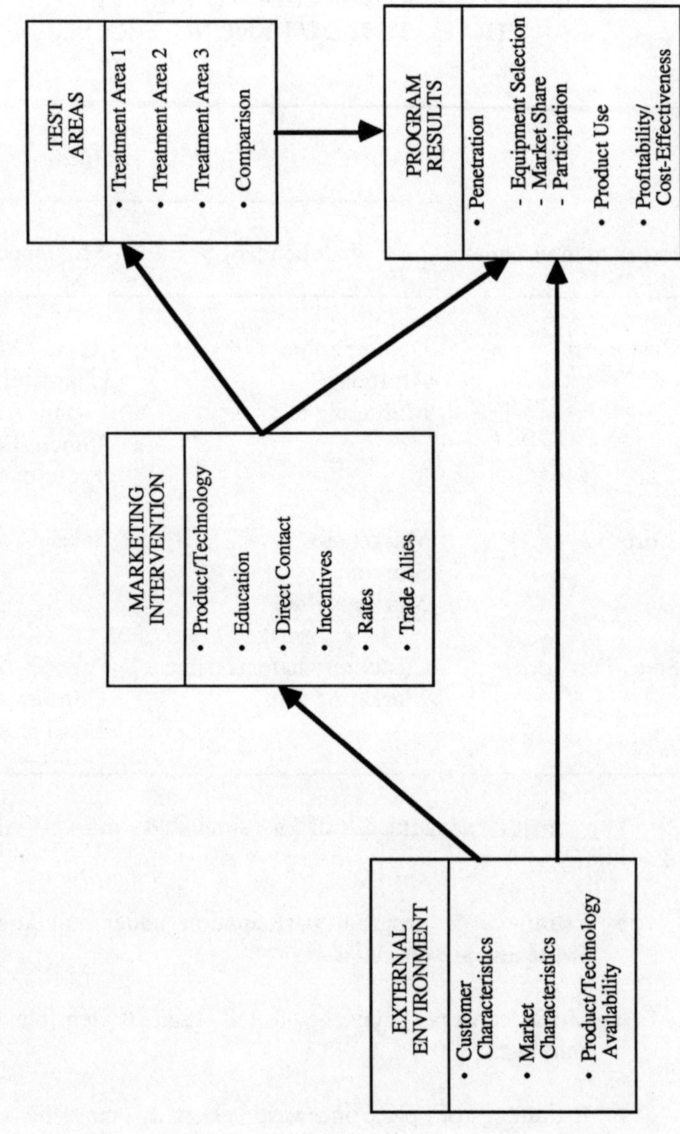

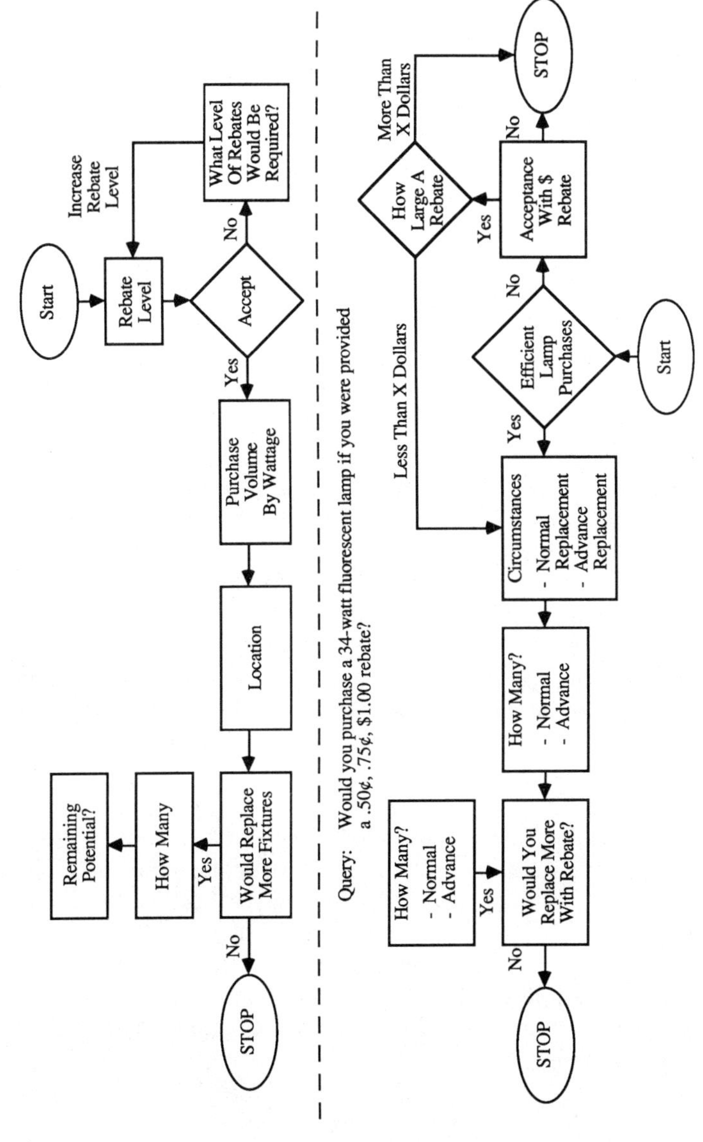

Figure 7-3
ILLUSTRATIVE SIMULATED TEST MARKETING DESIGN

At the end of a simulated market test, an S-shaped diffusion model can be used to estimate the impact of a full-scale program under various scenarios and economic conditions.

The largest cost elements of these programs over time are usually the variable cost components, which are the incentives. Consciously establishing desired reliability levels prior to field testing can result in significant cost savings, which may be a major consideration for utilities with very limited research budgets.

There may be occasions when a utility needs to establish a number of distinct test areas in order to control for normal market acceptance of the technology and the varying levels of incentives and promotions that are used. The actual number of test areas used depends on the number of marketing options being tested. For example, if a utility wanted to test the results of a marketing program containing three different program attributes, and also to adjust the results based on normal market acceptance, a design that utilizes four test populations would be required. As an extreme example, if a utility had four marketing program options to test, and if each option had four levels of attributes to test, a design that uses 16 groups would be required (i.e., 4x4=16). Realistically, this may be an impractical design to employ in the physical setting of a utility's service territory, unless the design is one employed in a customer trade-off study such as conjoint analysis.

The important point that needs to be made is that the test and comparison areas be as similar as possible; otherwise, problems related to external validity (meaning that generalizations from the test to the larger population are in error) may occur. A good way to evaluate test group similarity is to look at the mean group scores from census and prior utility surveys and see how similar the group areas are. Similar comparisons should also be made after participants have entered the program.

Another important consideration is the relative degree of isolation of the test areas from one another. There is a need to ensure that "cross-group" contamination does not occur from one test area to another. This problem may be totally unavoidable for smaller utilities serving relatively compact areas; a simulated market test may be more appropriate in such circumstances.

The next question that the utility planner may be asked to address is which research design is appropriate. There are a number of research designs that may be employed. A summary of the research designs available appears in Table 7-3. The latin square, randomized, and factorial are examples of such designs. Each of these designs has advantages and disadvantages. Some designs do a better job of testing for interactions than others. The problem, however, is that these designs require more observations per cell. Because the factorial design requires so many combinations, conjoint analysis may be the most appropriate technique to employ for reducing the range of program options. Once the conjoint is completed, the top two to four program combinations can be identified and a field test marketing experiment performed for these higher ranked options. A combination time series/cross-sectional design can then be used to experimentally determine market acceptance of the program options. Some designs require far more test areas than others, which may have a definite bearing on cost and challenge the limits of practicality.

Finally, some designs do a better job than others in controlling for historical and maturation bias. For example, the factorial design presented for a utility demand-side program in Table 7-4 shows that three levels of incentives, two types of promotions, and three different communication strategies are being tested.

Figure 7-4 provides an illustration of the combination time series/cross-sectional design. The basic principles of this design are the following:

- All groups are assumed to be equivalent.

- Benchmark measurements are taken of the participant/non-participant groups.

Table 7-3. SUMMARY OF SELECTED RESEARCH DESIGNS

Table 7-3
SUMMARY OF SELECTED RESEARCH DESIGNS

Design	Major Features	Weaknesses	Applicability to DSM Programs
Time Series	Test units are observed over time, both before and after exposure to program	Failure to control for history; useful for single-variable experimentation	Advertising and communications tracking and image tracking. Trade ally acceptance of utility DSM programs
Post-Test Only Design with Non-Equivalent Groups	The treatment occurs before pre-testing; secondary data may overcome this limitation	Absence of pre-test and self-selection bias. May result in non-equivalent groups	Measuring the effects of promotional programs involving dealers and other trade allies
Combination Time Series/Cross-Sectional Control Group	Time series measurements taken, as well as pre- and post-program measurements of participants and non-participants	No major weaknesses in this design, except for statistical and measurement bias which may apply to all other designs as well	Very useful approach for designing test marketing programs of DSM technologies. The "net" market share impacts and effectiveness of marketing strategies can be tested.
Completely Randomized Design	Simplest experiment; usually tests for single variable impacts. No matching according to external criteria. Analysis consists of average response of variable dependent	Limited control over extraneous factors. Useful for single-variable experimentation	Trade ally acceptance studies and advertising/communication effectiveness research
Latin-Square Design	Used when two important extraneous influences are controlled (e.g., time and price). A treatment or test variable occurs at least once in each block or test period. There have to be as many blocks as test periods	Usually requires a sizeable number of test groups, unless unbalanced design is used	Rate design and testing for financial incentive levels
Factorial Design	Useful in testing many decision variables that are highly interrelated. Customer acceptance may be a function of the combined marketing effort rather than the sum of individual marketing activities. Two or more marketing variables can be tested at the same time. Measures of interaction and independence can occur.	Costly to complete and requires many test areas, unless unbalanced designs are used	Optimizing the selection of DSM programs consisting of a bundle of marketing activities.

Table 7-4. FACTORIAL DESIGN FOR UTILITY DSM PROGRAM

Incentive/ Price	Promotion	Communication		
		X	Y	Z
Level 1	A	1-A-X	1-A-Y	1-A-Z
	B	1-B-X	1-B-Y	1-B-Z
Level 2	A	2-A-X	2-A-Y	2-A-Z
	B	2-B-X	2-B-Y	2-B-Z
Level 3	A	3-A-X	3-A-Y	3-A-Z
	B	3-B-X	3-B-Y	3-B-Z

Note: 18 combinations required to be tested (3 incentive options x 2 promotions x 3 communications options).

- Post-program measurements are taken.
- Marketing program effects are based on subtracting the post-group scores for the test areas from the non-participant scores for the comparison area, with both scores adjusted for any benchmark differences that may exist.

Table 7-5 provides an example of this design in terms of the mathematical adjustments that are made. The cross-comparison design may render a higher result than expected due to omissions in the adjustments to the original characteristics of the population groups. The time series design may underestimate program effects. The combination design provides a more reasonable mid-point estimate of program effects.

Figure 7-4. TEST MARKETING/EVALUATION DESIGN: MATCHED COMPARISON APPROACH

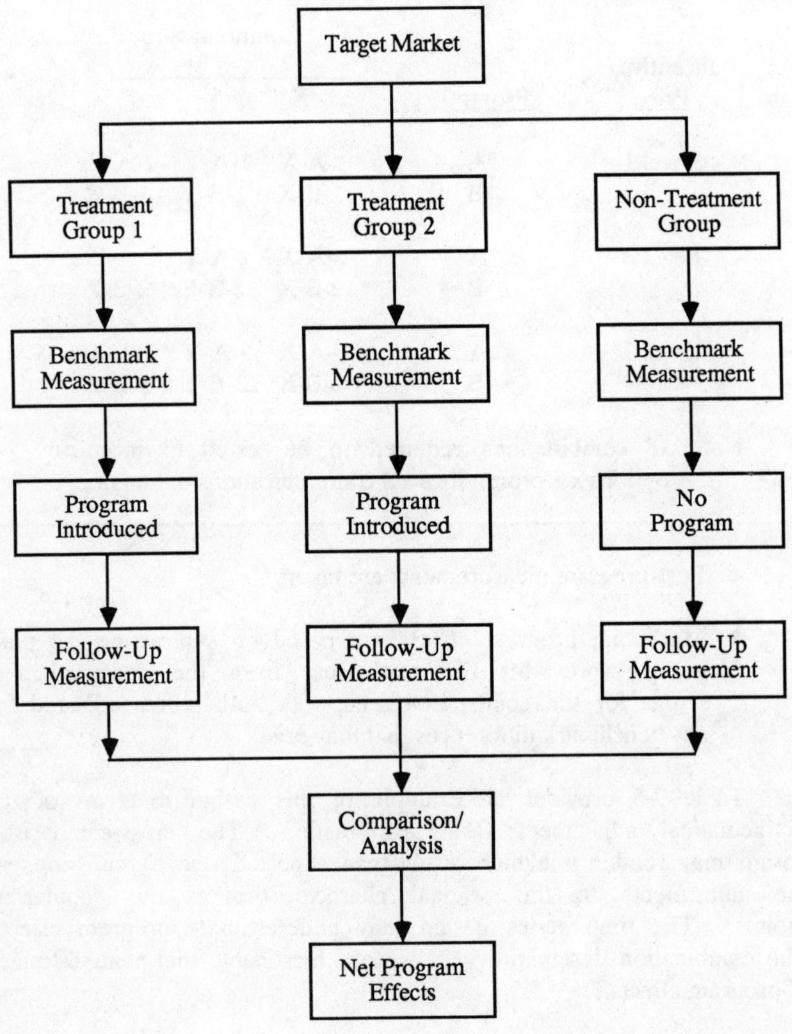

Table 7-6 shows the various statistics that are appropriate for use in a test marketing study. These statistics are usually available from application software programs that are routinely employed, and should be used to profile and evaluate the data that have been collected. The applicable test statistic for each of the various experimental designs presented above is also provided in Table 7-6. These methods are useful in determining whether the results from the study are statistically significant. Statistical significance is a measure of the likelihood that results occurred due to chance as opposed to being based on meaningful statistical relationships. The most common statistical techniques that are used are the t-test, F-test, and chi-square.

For the factorial and latin square designs, the F-test is most often used as the test of significance. This same statistic is used in analysis of variance and regression analysis.

Table 7-5. HYPOTHETICAL HEAT PUMP SALES FROM TEST MARKETING*

	Treatment Area	Comparison Area
Pre-Program Sales	100 (units)	125 (units)
Post-Program Sales	150 (units)	140 (units)
Effect of Marketing Program	(150-100) - (140-125) = 35 (units)	

*This table also shows how inflated utility estimates of program impact can be if all heat pump sales are considered (i.e., 150) or if the pre- and post-evaluations are only completed for the test area (i.e., 50). The table also shows how program effects are underestimated if a cross-comparison design is used with no benchmark measurement (i.e., 150-140 = 10).

Table 7-6. USING THE APPROPRIATE STATISTICS TO GAIN MEANING FROM TEST MARKETING DATA

Technique	Purpose
Standard Deviation	Useful to measure dispersion around group means.
Mean Deviation	Applicable when there is extreme skewness of data.
Coefficient of Variation	Determine which set of data is more or less variable if units are measured differently.
Estimated Standard Error of the Mean	Applicable where the standard deviation of the population is unknown for marketing problem solutions.
Confidence Intervals	Establishes importance of estimate.
t-Statistic	Used to compute significance of a statistically derived estimate for one or more samples. Mean scores of impact criteria are standardized.
F-Statistic	Statistic based on the ratios of variances between test and comparison groups.
Chi-square	Determines the dependence between two or more variables based sum of squares.

There are a number of bivariate and multivariate techniques that can be used in a test marketing program. It is beyond the scope of this chapter to fully describe their applicability to potential marketing programs. Table 7-7 is provided as a summary description of these techniques and their applicability to test marketing problems for those interested in an initial screening of the methods that are available. A hypothetical statistical analysis of a test marketing program is presented below as a case study utility test marketing problem.

Table 7-7. DATA ANALYSIS TECHNIQUES AND THEIR PURPOSES AND APPROPRIATENESS

Technique	Purpose	Appropriateness	Statistical Test
Analysis of Variance	Test for significant differences among average responses due to controlled variables	Analysis of categorical differences of groups (i.e., participants, non-participants). Dependent variable is interval.	F-test
Analysis of Covariance	Used in factorial designs where significant interactions among variables may occur.	Analysis of categorical differences of groups (i.e., participants, non-participants). Dependent variable is interval, especially when accounting for the interaction among independent variables.	F-test t-test
Correlation	Test for bivariate relationships between variables	Useful in measuring associations between variables.	Z-test based on sample correlation
Regression	Used to model association between dependent variables and a set of independent variables.	Useful in predicting responses and sensitivity analyses.	F-test
Discriminant Analysis	Used to classify population groups and describe predefined customer variables.	To identify group similarity/ differences and their significance.	Chi-square

CASE STUDY APPLICATION

The Research Problem

An electric utility was interested in developing a test marketing program to evaluate the relative effectiveness of communication-only versus rebates (i.e., $35 and $50) in stimulating purchases of more efficient appliances (as measured in terms of their annual energy consumption). The utility wanted to take into account a number of factors that might influence appliance purchase behavior, including prior sales ratios for each dealer, customer characteristics, and the relative effects of the programs. An explicit set of research objectives was formulated to address these issues. These objectives were stated as follows:

- Determine the relative effects of communication alone and in combination with other marketing elements.

- Determine the relative effects of the different rebate levels in terms of incremental increases in customer purchases of more efficient appliances

- Obtain relevant experience in the operation of the program on a pilot scale.

The Research Design

The research design employed called for three test areas and one comparison area which was to represent "normal" appliance purchase patterns. The programs implemented in the test areas were as follows:

- Area 1 -- communication only

- Area 2 -- communication and $35 rebate

- Area 3 -- communication and $50 rebate.

As a basis for the design of the program, a literature search of information on other utility programs was completed. The major characteristics of the programs and participation levels were identified. Dealer meetings and focus groups with utility customers were held to identify the key acceptance factors as they relate to the major product attributes and marketing strategies. This information was used to identify the positioning statements for dealers and customers. The type and levels of incentives were also finalized based on the results of the qualitative research.

The type of research design employed was a combination time series/cross-sectional comparison. It was felt that this design would adjust for the different appliance sales patterns by dealer more than any other design.

Next, the major research hypotheses were formulated. A few of the many hypotheses that were used to develop the study design included the following:

- There will be no significant difference in the demographic characteristics of participant and non-participant customers who purchase efficient versus non-efficient appliances, respectively, in the test areas.

- The higher the financial incentive that is offered, the greater the impact on the relative efficiency levels of appliances that are purchased.

- Rebate programs will be more influential than information programs alone in stimulating purchase of more efficient appliances.

Based on the research hypotheses, the expected relationships of the data that were used in the study can be tested. The primary data collection tools were:

- Rebate application form (demographics, attitudes, buying influences, and model size and type)

- Follow-up customer survey (buying influences, additional demographics, and information sources)

- Appliance sales tracking system (model numbers, brand names, and annual energy cost)

- AHAM list of appliances and model numbers.

In order to qualify for rebates, customers were required to submit rebate application forms along with the sales receipt and federal energy guide label. A follow-up survey was enclosed with the rebate checks to obtain additional customer background and decision-making information. Dealers were asked to keep track of total sales volumes. Dealers in the comparison area were also asked to provide names of appliance purchasers so that similar background information could be obtained through a follow-up survey.

Statistical Techniques

The statistical techniques employed included the following:

- Univariate statistics to measure central tendency and dispersion

- Simple correlation and regression in order to analyze the rebate application and follow-up survey data

- Difference of means test for proportions.

The F-test and t-tests were used to determine the relative significance of the correlation coefficients.

DEVELOPING THE PROGRAM

The program design process consisted of the following activities:

- Design and development of the necessary communication materials, including dealer kits

- Design and development of the necessary data collection materials

- Development of a monitoring and tracking system

- Development of an organized means to ensure that rebates are paid

- Development of an organized means to roll out the program.

The customer communication materials used consisted of radio spots, newspaper ads, direct mail, and point-of-purchase materials. Rebate application forms and follow-up surveys were also designed and printed. A coupon processing vendor was used to help process all the rebate applications that were received. A series of report outputs was designed using the Statistical Analysis System in order to track monthly participation levels. The reports that were generated could be used on mainframe or PC systems. A formal announcement and program roll-out campaign was also scheduled.

The program ran for about two years. The reason for the duration of the program was that much of the dealer's stock had already been acquired; thus, customers' opportunities to choose high efficiency appliances may have been restricted in the first year of the program. It was assumed that dealer product availability would be greater in the second year and that this would enable the effects of the program to be evaluated more accurately. It was also assumed that the market response to the program would be more accurately reflected in the second year because information about the technologies and the availability of rebates had been more widely diffused.

DATA ANALYSIS AND FINDINGS

The data analysis techniques listed above were used to test for each of the previously presented hypotheses. The following conclusions can be drawn based on the data that appear in the corresponding tables:

- The first hypothesis -- that the demographic characteristics do not significantly vary in terms of buyers of qualifying and non-qualifying appliances -- can be accepted (Table 7-8).

- The $50 rebate and communication program does result in a statistically significant reduction in energy usage.

- The $35 rebate and communication program does not result in significantly improved efficiency levels of appliances (Table 7-9).

Table 7-8. DEMOGRAPHIC CHARACTERISTICS BY TEST AND COMPARISON AREAS*

Demographic Characteristics	Participants (N = 1085)	Non-Participants (N = 1000)
Age (Head of Household)	50.9 years	49.4 years
Income	$31,053	$31,538
Education (Years Completed)	11.6	14

*Assumes that both participant and non-participant population variances are the same on all three characteristics.

Table 7-9. UNITARY ENERGY IMPACTS OF PROGRAM

District	Average Energy Consumption (kWh/yr) Pre-Program	Program	Savings
Advertising Only	1004.5	1066.6	-62.1
Advertising Plus $35 Rebate	1056.2	1018.2	38.0
Advertising Plus $50 Rebate	1113.1	1077.2	35.9

- The hypothesis stating that the advertising-only program results in significant improvement in appliance efficiency is rejected because of the high statistically significant consumption levels (Table 7-9).

A number of factors need to be considered in analyzing the data. The first concern that exists is whether the variables being used in the study measure distinct phenomena. Another concern pertains to investigating spurious relationships. While the results at an aggregate level may show few if any relationships, a disaggregated analysis (i.e., at the appliance size level and dealer store level) may reveal that a significant relationship exists. There should be an aggressive search for rival explanations. A more disaggregated analysis may be helpful in a search for such explanations.

REPORTING RESULTS

In writing the final report for a test marketing study, a few guidelines are suggested. First, the executive summary is perhaps the most important part of the report. The findings that are presented should be tied directly to the research objectives. The completed data analysis should be presented in sufficient detail to prove the major points presented in the executive summary. Great care should be taken to avoid overwhelming the reader. Appendices should be used to provide additional information that is useful but not essential.

As noted above, an important element of test marketing research is the search for rival explanations. To do this requires that at least two hypotheses are being tested. In addition, there is a need to use comparison as the basis for reaching conclusions regarding whether the marketing options are "netting out" normal market responses and variations due to customer characteristics or other "confounding" factors. There is also a need to be conscious of validity (i.e., that the variables measure what they purport to measure) and reliability (i.e., that the results upon which hypotheses are being judged are occurring more than just by chance). The concern for validity requires that multiple measures be employed in order to overcome any bias that may result from the data collection methods. External validity relates to how representative and random the study design is. Comparison groups are an important means of addressing problems of external validity; however, the participant and comparison groups must be equivalent. In order to improve the reliability of the results, the size and variation of participant and non-participant behavior response patterns must be significant enough to disprove the hypothesis that there is no significant difference between the test and comparison groups.

LESSONS OF EXPERIENCE IN TEST MARKETING

In reviewing the "lessons of experience" learned through the implementation of a utility test marketing program, a number of comments can be made regarding some of the problems that are likely to be encountered. Perhaps the area in which the utility manager is going to experience the greatest difficulty is in addressing the need for validity and reliability, largely due to the fact that there are a number of factors in the physical world that can threaten the experimental integrity of a project.

Another area of concern is the dealer; specifically, the behavior of the sales personnel in the program. Trade ally cooperation is critical in any experiment involving multiple decision-makers. In order to properly evaluate the results of a test marketing program using sales data, dealers must cooperate by providing those data. These important data will include a listing of customer names, and monthly sales by model and size. It may be difficult to obtain such data in the comparison areas and in the test areas where dealer incentives are lower. Much more time and effort will be required to maintain the integrity of the comparison area than the test areas. One should also be prepared to find that a significant percentage of product model numbers do not match (about 20 to 25%). Much time will be required to match the model numbers. The problem that arises is that some models are specially produced for either large volume chains or other buyers.

Care must also be taken to ensure that response bias does not occur in the follow-up surveys of non-participants. This population group is generally more difficult to get responses from. A small financial incentive or prize for returning a follow-up mail survey or responding to a telephone survey may help overcome this problem.

An important external validity problem is ensuring that the test areas are similar to the larger market that is likely to be targeted for the full-scale program. Moreover, the test areas themselves need to be reasonably similar (except of course in terms of the different marketing programs being tested). This is where the realities of the real world are unforgiving. While the ideal experimental conditions in the physical world may not be present, utilities must search for the most suitable test and control conditions in order to ensure that biased results do not occur. There is a likelihood that bias will be present in every test marketing experiment; the task is to find ways to keep it to a minimum.

Embedded within the context of a test marketing project there is likely to be a "mini-marketing" project; i.e., trade allies and marketing representatives will have to be sold on the merits of the test marketing project itself. It should be no surprise to learn that trade allies and marketing representatives are not overjoyed at the thought of having to support a test marketing project. If this project represents one of the first serious contacts with trade allies over the years, then an even harder sell may be required. This is largely because trade allies may be a bit skeptical as to how successful a utility marketing program may be. However, if the program is developed with much trade ally input and if the program is quickly perceived as a success, allies will strongly back the program and may even support continuation of the program. What trade allies look for is whether the program contains elements that they feel are both important to the customer and relatively painless to the allies. Higher incentive levels, expeditious payment of rebates to customers, and attractively produced advertising materials are what dealers look for. Dealers also like being contacted on a routine basis (i.e., once a month) in order to ensure that the program is running smoothly. The dealers kit can also be a quite useful and important part of the project.

A test marketing program provides ample opportunity for problem management specific to experimental design and tactical program implementation factors. The results of such an exercise can be invaluable in terms of identifying the relative benefits and costs of the program, and what works and does not work. Because test marketing may represent a major expenditure of resources, the utility planner should be sure that sufficient time has been devoted to screening out all of the options that are available. Lower-cost screening methods should be used to arrive at a program design that appears to be technically and economically sound. In essence, the test marketing program must be as close to the full-scale program as possible in terms of the design and marketing strategies being tested. The only difference is the scale of the program. A major premise put forward in this article is that there is much risk in jumping into such a program. Because so many choices are available and it is an exception for new products and services to be accepted in the market, the risks of moving immediately into a full-scale program are too great. There is the risk that customers may not be enticed by the program, or that customers may be significantly influenced by a program only if the program has too liberal an incentive (and thus is inefficient in obtaining customer participation). The task for the utility planner is to ensure that marketing program selection is based on measurable criteria that are as close to real market conditions as possible.

CHAPTER 8

Analyzing Market Penetration of New Technologies

Dilip R. Limaye and Craig McDonald

INTRODUCTION

"Build a better mousetrap and the world will beat a path to your door"; so goes the old cliche. But many inventors of "better mousetraps" have lost their fortunes because they did not address two crucial questions: Will this product be accepted by potential customers, and how much of the market will it penetrate? A key task faced by the utility DSM planner or marketing manager is predicting the market acceptance of the technology being promoted. The problem is particularly challenging in the commercial and industrial sectors because of the diversity of customer types and the complexity of the decision processes. This chapter describes the various approaches for analyzing market penetration of new technologies.

Predicting the market success of a new product or technology is one of the key problems for most companies, since business growth and success generally depend directly on product sales. Electric utilities have increasingly faced the problem of assessing the customer acceptance of new end-use technologies and utility programs designed to implement demand-side management (DSM) measures.[1] Many utilities have initiated efforts to promote products and technologies such as heat pumps and thermal storage, or adopted programs to implement conservation and load management. These utilities need to estimate the number of adopters or participants over time in order to assess the potential impacts on and benefits to the utility's long-range supply-demand balance. Utilities have a broad range of marketing methods available to influence the adoption or penetration of DSM options. These include various types of customer information, direct customer contact, advertising and promotion, alternative pricing, direct financial incentives, and trade ally programs.[2] It is very important for utilities to develop analytical tools to predict the market penetration of new technologies and the effect of various utility actions on that penetration.

METHODS FOR ANALYZING MARKET PENETRATION

The market acceptance or penetration of a product or technology can generally be viewed as a three-step process.[3] This process is illustrated in Figure 8-1. In the first step, customers form preferences regarding product attributes; these are sometimes referred to as "values." In the second step, the customers' preferences or values are translated into choices. Finally the choices interact with the dynamics of market behavior to determine the time path of acceptance or penetration.

Most methods for forecasting market penetration can generally be divided along these same lines. First, there are methods that estimate customer preferences (preference or value models), which are represented by scores or rankings of products and product characteristics. Second, there are methods that estimate market potential based on customer choice (choice models). Market potential, also called the equilibrium market share, refers to the fraction of consumers that would choose a product given a "level playing field" where biasing market forces are absent. Third, there are methods that forecast the actual sales of a product or the diffusion of a new technology over time (behavioral dynamics or diffusion models). Market diffusion refers to the pattern by which the sales potential is (or is not) achieved over time. Applying each of these methods provides useful information necessary to forecast market penetration. However, a true dynamic market forecast results only when all three methods are integrated.

Figure 8-1
PROCESS OF MARKET ACCEPTANCE OR PENETRATION

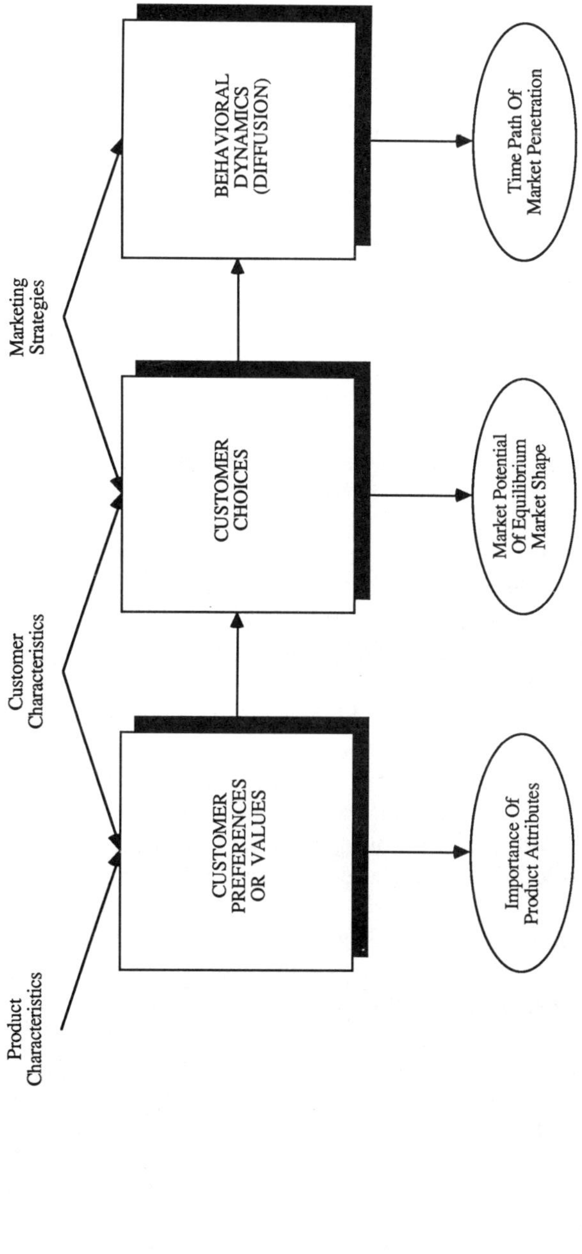

Product Characteristics

Customer Characteristics

Marketing Strategies

CUSTOMER PREFERENCES OR VALUES

CUSTOMER CHOICES

BEHAVIORAL DYNAMICS (DIFFUSION)

Importance Of Product Attributes

Market Potential Of Equilibrium Market Shape

Time Path Of Market Penetration

Figure 8-2 illustrates the different types of models. Table 8-1 provides a comparison of some of their characteristics. A brief discussion of each of these types of model structures is provided below.

Preference or Value Models

Methods that estimate customer preferences toward product attributes are generally called preference or "value" models because they reflect consumer values. These models typically quantify customers' values for individual attributes of products, and then combine these values to score or rank products. Assuming that customers maximize their value or "utility," these models can be used to predict which product(s) consumers would choose.

Value models are generally based on historical data on purchase patterns, or on market surveys. In some surveys, the consumer is simply asked directly about preferences. In other surveys, a variety of tradeoff questions about product characteristics are asked. Differences in methods of collecting and interpreting data form the basis for competing approaches to value modeling. Some of the major approaches to value modeling are discussed briefly below.[4]

- **Structural Cost Models** -- The simplest approach to value modeling is to assume that preferences are based solely on economic impact. Clearly, this approach can only be applied to products where cost plays a large role in market behavior. With such an approach, a structural model of each consumer's acquisition and use of the product is used to calculate the cost (e.g., net present value) to the consumer. Each product can then be ranked on its impact relative to other products, considering only economic issues. These models are often a starting point for more thorough analyses.

Figure 8-2
MARKET PENETRATION ANALYSIS METHODS

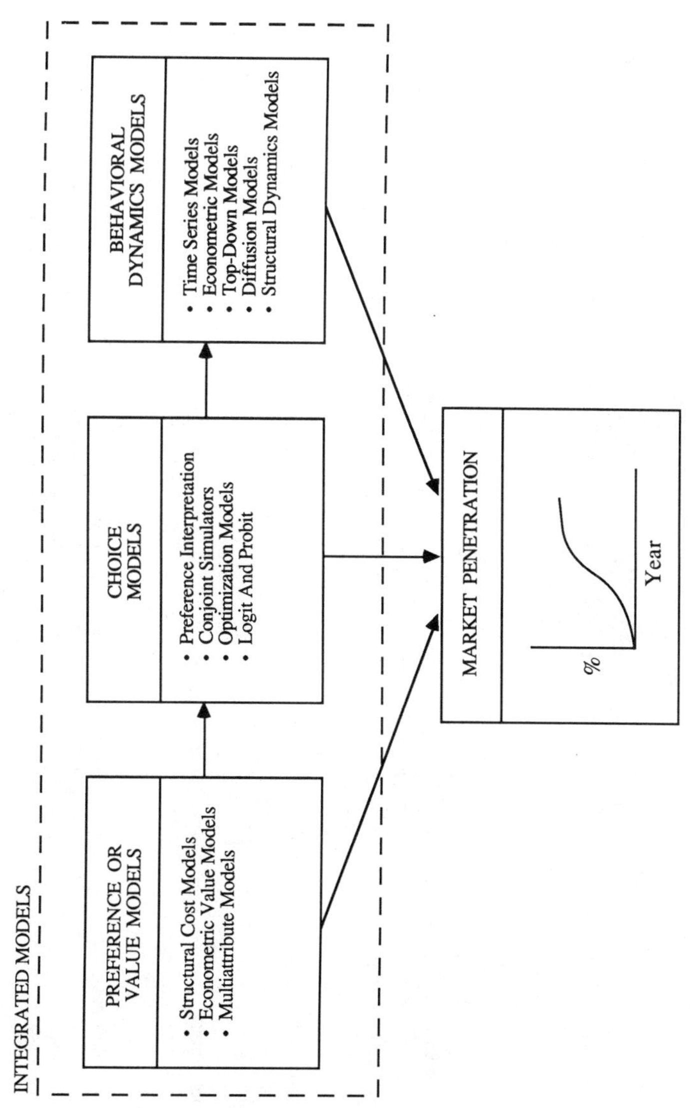

Table 8-1
A COMPARISON OF THE THREE METHODS FOR ANALYZING MARKET PENETRATION

Characteristic	Preference or Value Models	Choice Models	Behavioral Dynamics Models
Individual decision-making process	Capture only part of the process	Capture complete process	Represent time path; some models can capture complete process
Time aspects	Values are subject to change over time due to a variety of uncontrollable influences	Do not capture time dependent aspects	Explicitly model penetration over time
Calibration	Very sensitive to data collection instruments	Difficult to calibrate accurately unless sufficient individual-level data are available	Calibration based on historical data for similar products
Other comments	Not particularly suitable for new innovations	Do not usually account for external influences like availability, sales force efficiency, etc.	Some models are particularly good in that they account for external influences like price/advertising/subsidies, etc.

- **Econometric Value Models** -- In an econometric approach, consumers are divided into market segments. In each segment, the fraction that purchases a particular product is assumed to be a function of the product attributes and the market segment demographics. Parameters for such models are estimated statistically for each segment to best fit the sales data. With such a model, products are ranked based on the aggregate values attached to attributes in each segment. These models are widely used, particularly for traditional products in established markets.

- **Multiattribute Models** -- In a multiattribute approach, each consumer's value for a product is assumed to be a function of the value he attaches to individual product attributes. With data from surveys of attitudes toward product attributes, consumer values for existing or proposed products can be calculated. These models are gaining increased acceptance for analysis of markets for new products and services.

Choice Models

Methods that estimate market potential are generally called "choice" models because they reflect the choices that customers would make among different products. These models typically take information on how consumers value products (perhaps from a value model) and predict purchase behavior in a kind of vacuum, where no "market forces" affect consumer behavior. Market forces include considerations such as awareness of new products, sales force effectiveness, wearout rate of the existing stock, etc. In short, the market potential of a product is the share that product would capture if product attributes were the only things that influenced purchasing behavior. Although the market potential of a product does not directly predict sales, it is a key input to such an estimate.

Once a view of consumer values has been formed, there are two basic options for a choice model: deterministic or probabilistic. In a deterministic choice model, estimated values are assumed to reflect consumer preferences precisely. Consequently, the choice model indicates that each consumer will choose the product with the highest calculated value. Probabilistic choice models calculate the probability that a consumer will choose a given product. Of course, the probability that a particular product will be chosen rises as its value rises relative to competing products. However, choices remain uncertain because of heterogeneity among consumers, modeling simplifications, consumer inconsistency, and the like. Some of the most widely used probabilistic choice models including logit, multinominal logit, probit, and multinominal probit models.[4]

Behavioral Dynamics Models

Methods that forecast the pattern of a product's sales over time are often called "dynamics" models or sales forecasting models. Some dynamics models focus on the adoption of new products, and others deal with forecasting future sales of an existing product. These models may explicitly take into account the interaction of market forces or extrapolate from past trends and market indicators.[4]

Forecasting a product's sales over time is a difficult task because there are many factors involved, and the relationships among the factors and between these factors and sales are highly uncertain and not easily quantifiable. Some of the factors which must be considered when forecasting sales are:

- Inertia (reluctance of consumers to change from established products)
- Customer awareness of product
- Competing products
- Wearout rate of durables
- Advertising
- Sales force effectiveness

- Production constraints

- Distribution channels.

Some examples of behavioral dynamics models are described briefly below.

- **Econometric and Time Series Models** -- Econometric and time series models base market penetration on historical data. Time series approaches use trend extrapolation, regression, and autoregression techniques to predict future sales forecasts and make statistical correlations of sales with independent factors that influence sales, such as product price, price of competing products, disposable income, etc.

- **Diffusion Models** -- Market diffusion models, first introduced by Bass in 1969, are used to describe the introduction of a new technology into a market. Total sales or market share are assumed to follow a specified functional form (usually an S-shaped curve). The logic behind this assumption is that sales grow slowly at first because consumers are unaware of the new product, then accelerate, and level off as the potential market becomes saturated. The parameters of the function are supplied by historical (and often industry-specific) data.

- **Structural Dynamics Models** -- Using a model of the customer decision process (or some measure of market potential), structural dynamics models depict how market forces (such as wearout rate of durables, customers' reluctance to change, customer awareness, effectiveness of sales force, product attributes, and prices of competing products) influence the sales or market share of a new or existing product. Because structural models explicitly address the customer decision process and relevant market forces, they can lend significant insight into the market.

Integrated Models

Integrated market models combine the functions of the preference, choice, and behavioral dynamics models. They forecast product sales by considering all three of the key determinants of market behavior: how customers value competing products, how customer values affect their preferences, and how market forces affect penetration over time.

MODELING THE MARKET PENETRATION OF NEW ENERGY TECHNOLOGIES

Models of energy technology choice have been evolving steadily over time.[4] The original models were strictly econometric. The econometric models suffer from several major limitations, including:

- Their performance in predicting fuel choice has been poor, particularly when limited data are available to estimate the model parameters.

- They are limited in their ability to analyze the effects of programs, standards, and changes in technology characteristics.

- They cannot be used to analyze the penetration of different technologies that use the same energy source, such as electric resistance heaters and heat pumps.

Engineering-economic models were developed to overcome the shortcomings of the econometric models. The engineering-economic models generally assume that the option with the minimum life-cycle cost would be selected. One difficulty with this approach is the choice of the appropriate discount rate for the decision-makers. A major limitation is that the engineering-economic models imply that all decision-makers would select the same system, yet we never observe all building owners selecting the same fuel and equipment.

A new class of models that is more useful for analyzing the commercial sector energy technology choices is the microsimulation approach using a discrete choice model. This approach determines the technology choice that will yield the minimum life-cycle costs for each of a sample of decision-makers. Each decision-maker is represented as having a unique discount rate and different expectations about future fuel price escalation. Decision-makers can be partitioned into different groups representing differences in factors affecting investment decisions, such as ownership status (for example, owner-occupants, developers, and government decision-makers). This approach has been incorporated into a model developed by Synergic Resources Corporation called COMPEN.[5] This model has been used by utilities in the Pacific Northwest, Florida, North Carolina, California, New Jersey, and Pennsylvania to forecast the penetration of commercial end-use technology options such as high efficiency heat pumps, thermal storage, cogeneration, etc.[6,7,8,9]

The basic modeling approach, shown in Figure 8-3, is to represent the technology choices made by the various decision-makers by using a Monte Carlo simulation approach. The simulation addresses the specific choices by discretely analyzing the available fuel/technology combinations. The major inputs to the discrete choice analysis are:

- Characteristics of the decision-makers, including their discount rates, fuel price escalation expectations, knowledge/perceptions of technologies, and attitudes/ preferences

- Characteristics of the competing technologies, including capital costs, operating and maintenance costs, performance, reliability, maturity, etc.

- Other information, such as economic ground rules (inflation rate, interest rate, etc.), typical building characteristics (square feet, number of stories, design features, etc.), regulations, building codes, tax credits, etc.

Figure 8-3. CONCEPTUAL OVERVIEW OF DISCRETE CHOICE SIMULATION MODEL

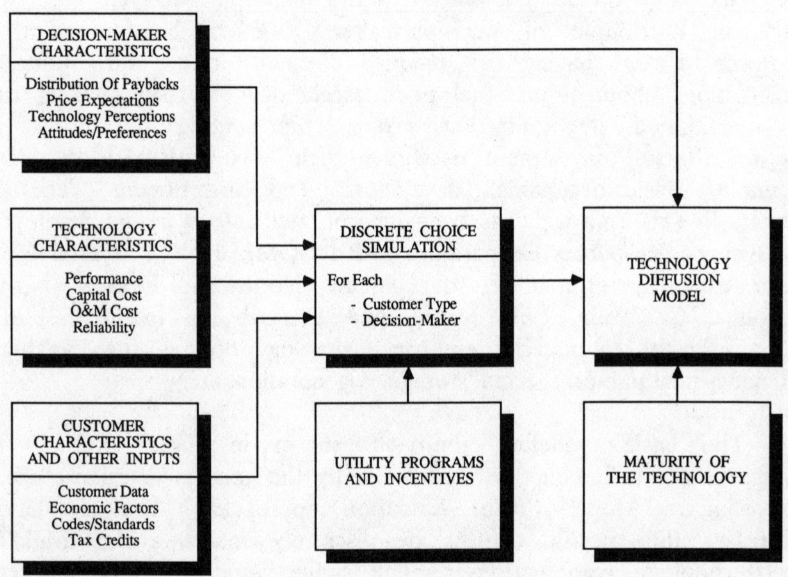

- **Utility programs and incentives.**

The information regarding decision-maker characteristics is represented as probability distributions to capture the variation among the different individuals within a decision-making group (such as builders, architect-engineers, etc.). These distributions include discount rates, fuel price expectations, and technology perceptions. Similar distributions can also be used for some technology characteristics, particularly perceived reliability, capital costs, etc. These distributions influence the variability of the decisions, which is represented in the Monte Carlo simulation by choosing randomly from the distributions.

The simulation is performed separately for each group of decision-makers for each building type. Data on the decision-maker characteristics and perceptions are developed from surveys. Similarly, survey data are used to determine the degree of influence (weighting factor) each decision-making group has over fuel/technology choice decisions for each building type.

The discrete-choice simulation estimates the equilibrium market shares for the different technology choices. A diffusion model can then be applied to predict the time path of technology adoption. A discussion of diffusion modeling is presented below.

A key feature of the discrete choice simulation approach is the simplicity in its representation of the fuel/technology choice decisions as they are actually made. By obtaining the appropriate data to develop the relevant distributions, realistic representations can be generated. By conducting surveys of decision-makers, these distributions can be refined and modified to further improve the performance of the model. An advantage of the simulation approach (over econometric methods) is that, by providing data on the discrete choices made by each simulated decision-maker, a greater understanding of the importance of various technology characteristics and decision-maker attributes is obtained, providing information useful for the design of utility programs and incentives.

Another major advantage of this type of model structure is its adaptability to the utility service area level. By developing appropriate inputs, the model can be easily transferred to different geographic areas. For example, typical customer characteristics can be developed for the relevant geographic area, and decision-makers in the area surveyed to develop the input distributions.

AN OVERVIEW OF DIFFUSION MODELING

The modeling of the process by which an innovation (technological or otherwise) is transmitted to, and subsequently adopted by, any desired segment of society has long been a multidisciplinary endeavor, spanning such areas as marketing, biology, economics, and forecasting. The underlying behavioral representation in all models is universal: An innovation is first adopted by a select few individuals who influence others, over a period of time, to adopt it. Adoption, therefore, is postulated to be a two-stage process: innovation by a few, and imitation of the few by many.[10] It is this imitative interaction indulged in by the majority of the target population which is the primary reason for the rapid growth state in the diffusion process. One characteristic by which one classifies a diffusion model is the explicitness of its formulation in including an expression for innovation, in addition to one for imitation.

The focus of all diffusion models is always on the generation of the first purchase sales volume curve of the innovation. Toward this objective, all diffusion models implicitly assume that the basic shape of such a curve is either exponential or S-shaped -- implying an acceptance of the product life-cycle concept and its inherent assertion that sales of an innovation will necessarily peak (reach "maturity"), regardless of marketing effort. Consequently, a second characteristic by which one classifies diffusion models is their ability to provide a shape that accommodates a flexible point at which sales may peak, depending upon the particular product market characteristics. A related characteristic is one of symmetry. Obviously, a curve which can handle both symmetric and non-symmetric sales patterns is a more versatile model.

The best known diffusion models of new product acceptance in marketing include the models of Bass[11] and Mansfield.[12] The Bass model and its revised forms have been successfully demonstrated in retail service, industrial technology, agriculture, educational, and consumer durables markets. The Mansfield model and its revised forms, such as those proposed by Blackman,[13] Fisher and Pry,[14] and Sharif and Kabir,[15] have been used in technological substitution studies of industrial innovations.

The basic structure of a diffusion model, as postulated by Bass,[11] is provided below:

$$\frac{dN(t)}{dt} = [\,p + q\frac{N(t)}{m}\,][\,m - N(t)\,]$$

where

N(t)	=	Number of adopters at time t
m	=	Maximum number of adopters (equilibrium market share)
p	=	Coefficient of innovation
q	=	Coefficient of imitation

A number of variations of this basic structure have been proposed and empirically tested. A recent study by Rao[16] compared the performance of five diffusion models in predicting the sales of four appliances: room air conditioners, dishwashers, clothes dryers, and color TVs. Table 8-2 shows the results of the evaluation.

A useful formulation of diffusion modeling was developed by Lawrence and Lawton,[17] based on empirical analysis of sales data on a number of consumer goods and commercial products. The model structure, shown below, was found to apply to a wide range of these products:

$$N(t) = \frac{m + N(o)}{1 + [N(t)/N(o)]e^{-p^{*}t}} - N(o)$$

where

N(t)	=	Number of adopters at time t
N(o)	=	Number of prior users (adopters at time o)
m	=	Maximum number of adopters
p*	=	Adoption rate parameter

The empirical studies for a number of consumer products yielded estimates of p* between 0.40 and 0.55 (with an average of 0.5). The value of p* for commercial products ranged from 0.65 to 0.68 (average of 0.66).

Table 8-2
A COMPARATIVE EVALUATION OF FIVE MAJOR DIFFUSION MODELS

				Rank in Terms of Fit to Real Data				Rank in Terms of Sales Forecast Accuracy			
Model	Innovation/ Imitation	Flexible Turning Point	Symmetry	Room Air Conditioner	Dish Washer	Clothes Dryer	Color TV	Room Air Conditioner	Dish Washer	Clothes Dryer	Color TV
Mansfield (1961)	Imitation only	No	Symmetric	4	5	3	4	5	5	5	4
Floyd (1962)	Imitation only	No	Non-symmetric	2	2	2	2	3	4	1	5
Martino (1975)	Imitation only	No	Non-symmetric	1	1	1	1	2	2	2	2
Bass (1969)	Both	Yes	Symmetric	5	4	5	5	4	3	4	3
NUI (1983)	Both	Yes	Symmetric or non-symmetric	3	3	4	3	1	1	3	1

Source: Sanjaykumar Rao[16]

ASSESSING EFFECTS OF UTILITY PROGRAMS AND INCENTIVES ON MARKET PENETRATION

A key problem faced by electric utility planners is the assessment of the potential benefits of utility programs and incentives on the penetration of a new end-use technology. Utilities have adopted a wide range of programs to promote technologies such as heat pumps, thermal energy storage, energy-efficient appliances, and industrial electrotechnologies.[18,19,20] Table 8-3 shows examples of different types of programs/incentives and their effects on market penetration. These different programs are described below:

- **Education and Information Programs**, such as bill stuffers, workshops, seminars, etc., increase the knowledge and awareness of the technology and therefore accelerate the diffusion process. This effect can be modeled by changing the parameters of the diffusion model.

- **Rebates** reduce the first cost of the investment in a new technology, thereby improving the economic attractiveness (payback or return on investment). This effect can be addressed in the discrete choice model by reducing the first cost.

- **Low-Interest Loans** reduce the debt costs and improve the economic attractiveness. This effect can also be addressed in the discrete choice model.

- **Third-Party Financing** can be attractive for customers who are constrained by capital availability. Surveys can be used to identify the potential market for this type of program.

- **Changes in Rate Structures** can be used to reduce the operating costs of a new technology. Rate structures can be modeled in the discrete choice simulation.

Table 8-3. EFFECTS OF UTILITY PROGRAMS/ INCENTIVES ON MARKET PENETRATION

Program/Incentive	Effect on Market Penetration
Education/Information	Accelerate diffusion
Rebates	Reduce first cost
Low-interest loans	Reduce debt repayment
Third-party financing	Eliminate capital requirement
Rate structures	Reduce operating costs
Design assistance	Accelerate diffusion

- **Design Assistance** (to architects and engineers) can be effective in accelerating the diffusion of the new technology. This effect may be addressed in the diffusion model.

An example of a utility application of a discrete choice simulation model combined with a diffusion model is given below.

ASSESSING MARKET PENETRATION OF THERMAL STORAGE: A CASE STUDY

Objectives and Approach

The North Carolina Alternative Energy Corporation (AEC), a research and development agency representing the electric utilities in North Carolina, performed an evaluation of the market potential for thermal storage in commercial buildings in North Carolina.[8] The approach used in this study consisted of the following steps.

- Identification of the commercial building types that are most suitable for thermal storage application

- Development of a prototypical building for each building type

- Modeling of the cooling loads for each prototypical building using an hourly load simulation model

- Analysis of the performance of conventional cooling systems and thermal storage in meeting the cooling loads

- Calculation of the capital costs of conventional and thermal storage systems

- Calculation of the operating costs of conventional and thermal storage systems using appropriate utility rate structures

- Estimation of market penetration with no utility programs, and with alternative rate structures and incentives.

Market Penetration Analysis Methodology

Figure 8-4 shows an overview of the structure of the market penetration model. The model consists of three major parts:

- Defining the size of the base market

- Identifying the subset of the market which is:

  -- purchasing HVAC systems
  -- informed about thermal storage
  -- willing to consider installing thermal storage

- Estimating the portion of the "willing" market for which thermal storage is cost-effective.

Figure 8-4. MARKET PENETRATION MODEL OVERVIEW

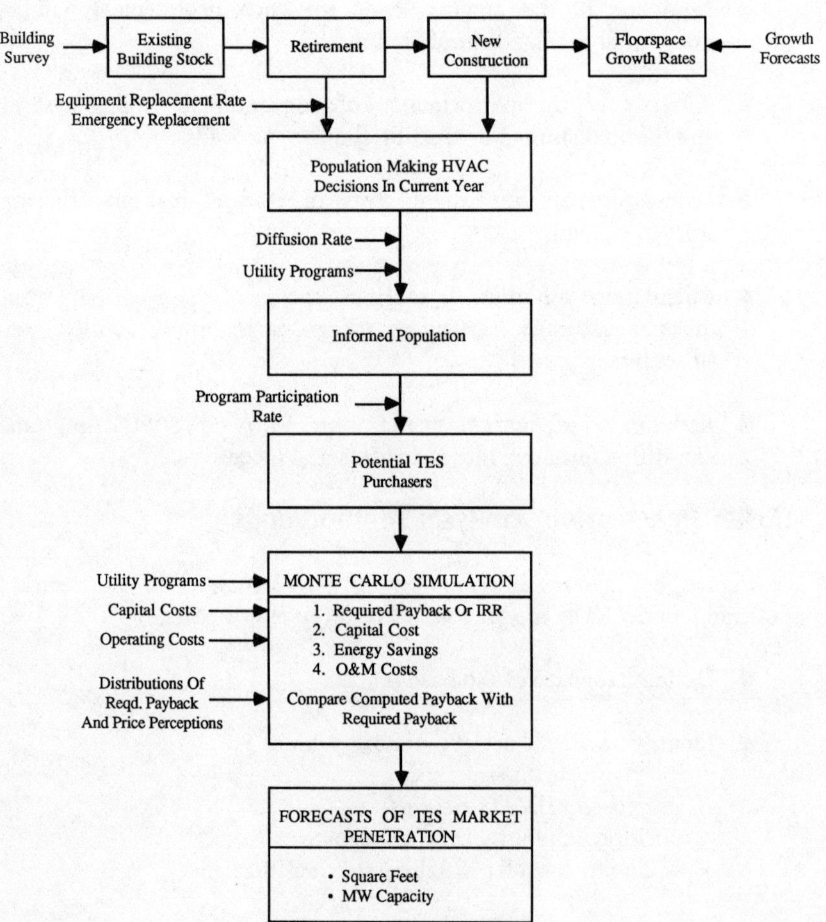

Defining the Base Market

Estimates of the 1987 floorspace stock were obtained from prior studies conducted by AEC. The retirement of building stock was then estimated using average building lifetimes. The percentage of the existing building floorspace in which the HVAC equipment is replaced in any given year was then calculated as a function of three factors:

- The initial distribution of ages of HVAC equipment
- The average lifetime of HVAC equipment
- The proportion of HVAC equipment that is replaced under emergency conditions.

It was assumed that only those who plan to replace their HVAC equipment were candidates for installing thermal storage systems in existing buildings. The new construction market is estimated as the sum of the growth in floorspace plus the retirements of existing floorspace stock.

Identifying Potential Thermal Storage Purchasers

The above calculations provide estimates of the total market for new (or rehabilitated) HVAC systems in each year. Only a portion of this market will consider thermal storage as an option because some decision-makers are either unaware or unconvinced of its potential benefits. The portion of the population that is aware of the potential benefits of thermal storage is called the "informed population." The size of the informed population depends on the number of prior purchasers. In each year, the portion of informed HVAC purchasers increases at a rate described by the diffusion function.

The Lawrence and Lawton diffusion model (described above) was used to estimate the informed population. The diffusion parameter in the model was assumed to be 0.66, based on the empirical studies of

Lawrence and Lawton for commercial products.[17] Utility information programs can affect the rate at which people become informed. It was assumed that the effect of information is the same as increasing the effective number of prior purchasers. Based on market research studies, information programs increase the market size by about 3 to 5%.[21] The effect of this method of modeling information programs is shown in Figure 8-5.

Only a portion of the informed population will participate in utility programs. It was assumed that the participation rates will be equal to the number of those who responded (in a telephone survey) that they would be somewhat or very interested in participating in each program option. These participation rates are shown in Table 8-4.

Summary of Low and High Scenario Assumptions

Calculations were performed for two scenarios, labeled as "low" and "high." Three major quantitative factors differentiate the low and high scenarios:

- **HVAC Replacement** -- The parameter which represents the fraction of HVAC systems replaced under emergency conditions, assumed to be 30% (.3) in the low scenario and 10% (.1) in the high scenario

- **Prior Purchasers** -- The estimated beginning percentage of prior purchasers obtained from the results of a phone survey and assumed to be 1% in the low scenario and 10% in the high scenario

- **Cost Effectiveness** -- The percentage of situations in which thermal storage is cost-effective for those that buy it, assumed to be 50% in the low scenario and 100% in the high scenario.

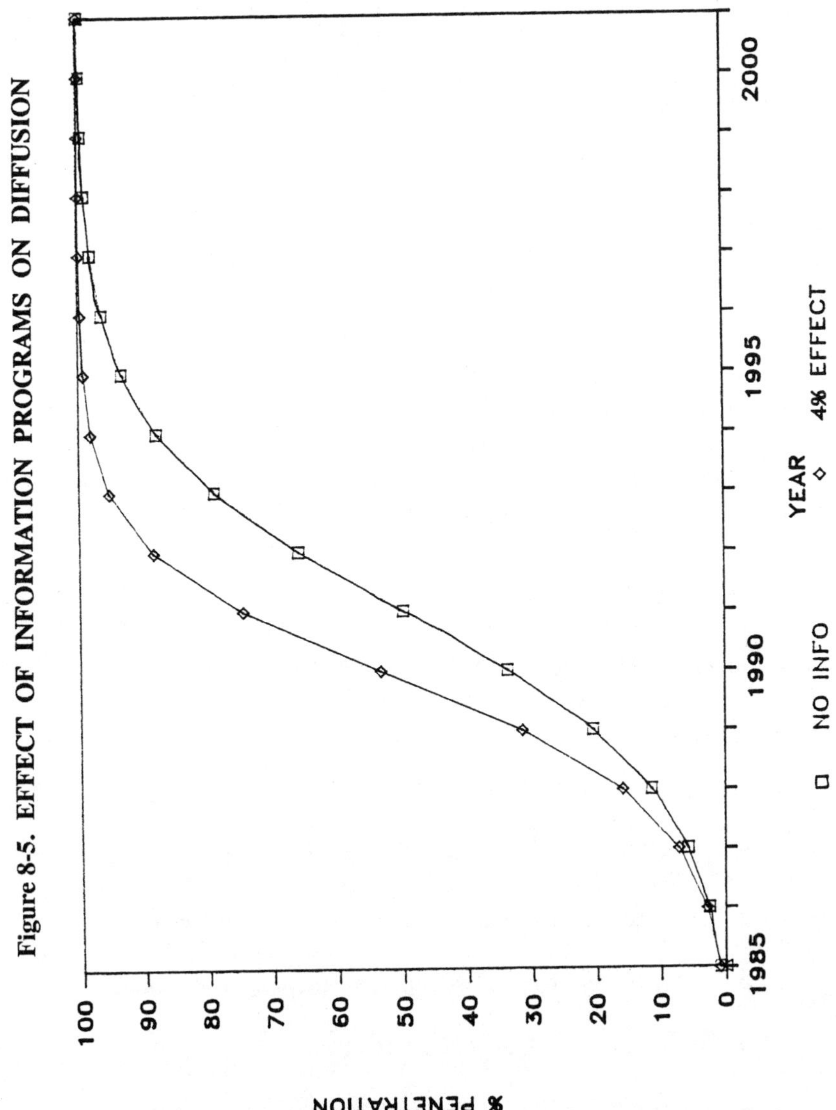

Figure 8-5. EFFECT OF INFORMATION PROGRAMS ON DIFFUSION

Table 8-4. EXPECTED PARTICIPATION RATES IN THERMAL STORAGE INCENTIVE PROGRAMS

Type of Program	Percent Willing to Participate
Guaranteed Payback	97
Guarantee of System Performance	92
Cash Rebates	83
Shared Savings Plan	78
Low-Interest Financing	51
Payment of Design Fees	81
Third-Party Financing	35

Source: Telephone survey of building owners and managers in North Carolina[8]

Estimating Market Potential

There are four market segments with differing characteristics:

- Existing buildings willing to participate in programs
- Existing buildings not willing to participate in programs
- New buildings willing to participate in programs
- New buildings not willing to participate in programs.

Each of these groups has a different set of characteristics. The discrete choice simulation uses a Monte Carlo technique to determine the percentage of the population within each group for which thermal storage is cost-effective.

Three sources of uncertainty were addressed in the Monte Carlo simulation:

- **Required Payback or Rate of Return** -- The distribution was defined from the results obtained through a telephone survey of commercial building owners. An illustration of the cumulative distribution of required paybacks from the survey for selected building types can be seen in Figure 6-2 in Chapter 6.

- **Capital Cost** -- A uniform distribution over the range of plus and minus 15% of the estimated capital cost was used to represent different site conditions.

- **Electricity Bill Savings** -- A uniform distribution over the range of plus and minus 20% of the estimated savings was used to represent the different price expectations of different decision-makers.

The output of the market penetration model provided a forecast of the square footage of space for which thermal storage systems will be chosen. This forecast was provided for both new and existing buildings for each of the representative building types. Using the results of the building simulation analysis, the square footage estimates can be translated into megawatts of load deferred.

Effects of Utility Programs

The major utility programs (to promote thermal storage) addressed in the study included alternative rate structures and financial incentives. Utilities can change the rate structure as a means of improving the economics of thermal storage. The rate structure is used to determine the operating costs. The following rate programs can be analyzed:

- Changes in the definition of the time periods for peak versus off-peak

- Changes in demand charges for either peak and off-peak periods or both

- Changes in energy charges for either peak and off-peak periods or both

- Use of a separate rate for the thermal storage system.

The model has the capability of analyzing four incentive programs:

- Rebates

- Low-interest loans

- Performance guarantees

- Guaranteed payback.

Each of these programs is analyzed by modifying the economic calculations in the Monte Carlo simulation for the subgroup of program participants. The modifications to the economic calculations and required input data to describe each program option are summarized in Table 8-5. For the guaranteed payback program, the payback of the thermal storage system must satisfy both the utility criteria (from input) and the decision-maker criteria (drawn randomly from the input distribution of required paybacks).

Table 8-5. APPROACH FOR ANALYZING INCENTIVE PROGRAM EFFECTS

Program	Strategy	Inputs
Rebates	Reduce capital cost by rebate amount	$/kW incentive
Low-Interest Loans	Add annual interest savings to electricity savings	Nominal interest rate and % points subsidy
Performance Guarantee	Increase required payback criteria	Years increase in required payback
Guaranteed Payback	Reduce capital cost until target payback is achieved	Target payback

Results

Figure 8-6 shows the estimates of market penetration of thermal storage (in terms of MW deferred) in the service area of Carolina Power and Light, one of the three major utilities in North Carolina, for the low and high scenarios. The estimates are shown for the "no programs" case and with rebates (of $200/kW deferred). The breakdown of the total MW by building type is shown in Figure 8-7 for the high scenario without any programs. The combined effect of rate structures and rebates for one building type is shown in Figure 8-8.

Figure 8-6
TOTAL MW DEFERRED BY THERMAL STORAGE, CAROLINA POWER AND LIGHT

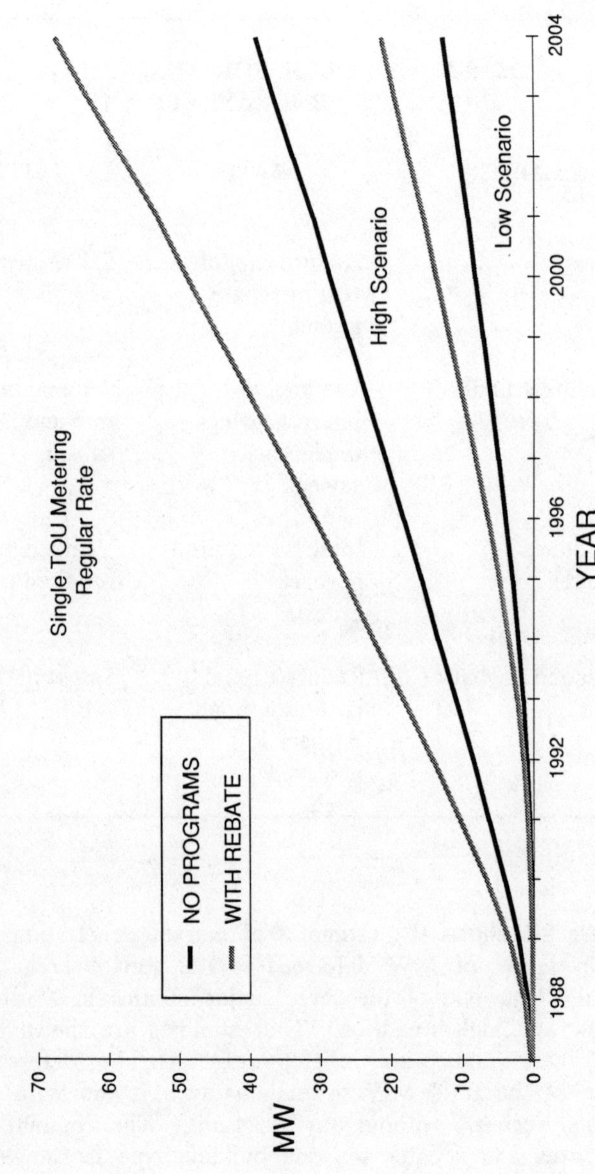

Figure 8-7
MW DEFERRED BY THERMAL STORAGE IN YEAR 2004 ACCORDING TO BUILDING TYPE, CAROLINA POWER AND LIGHT

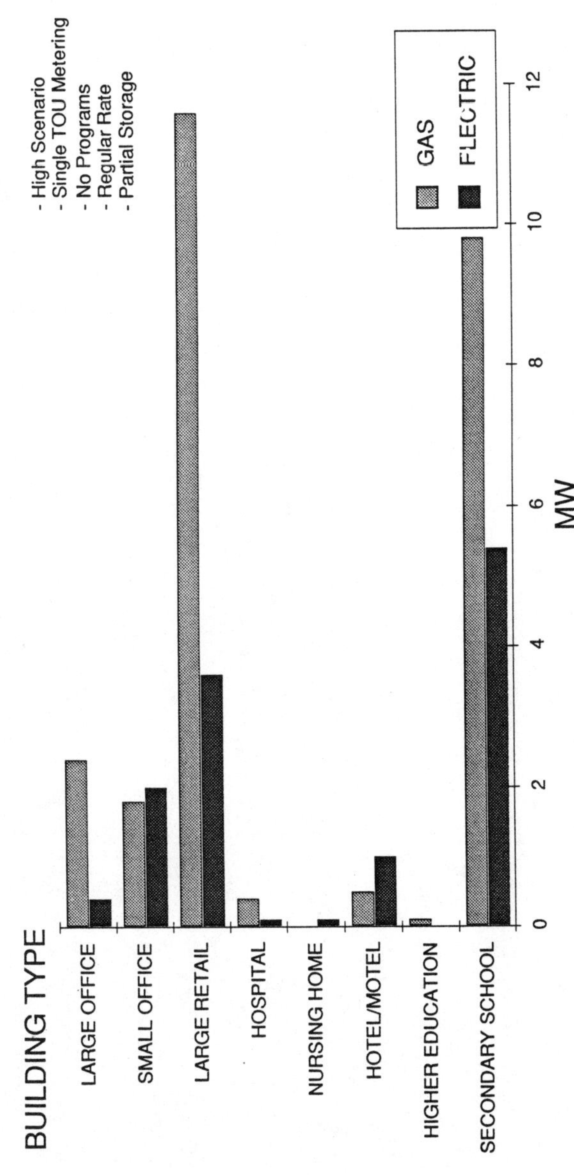

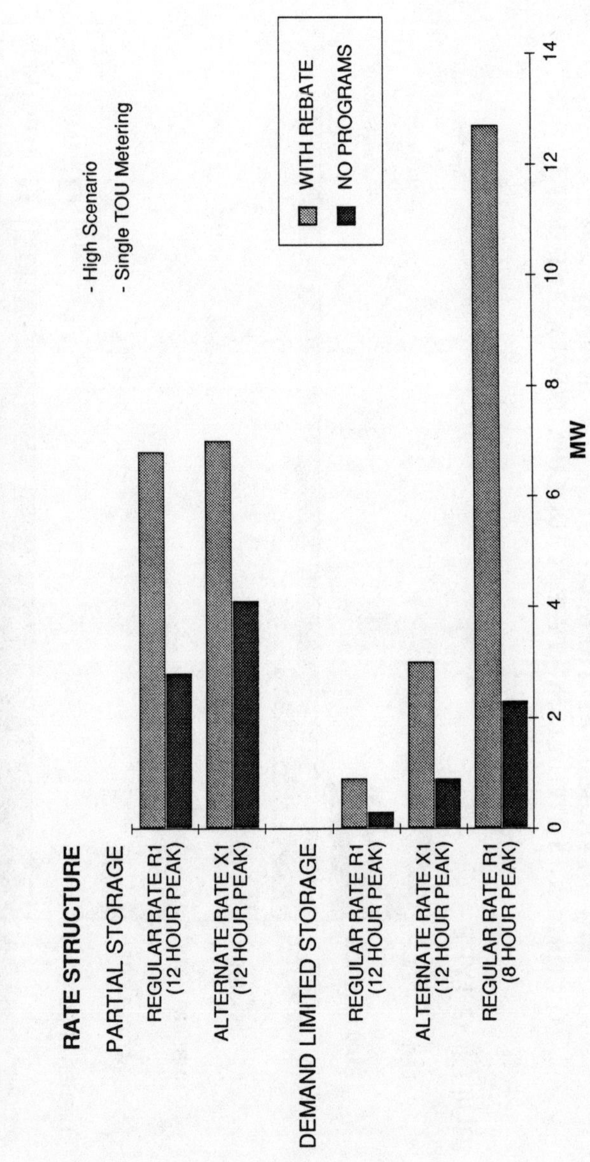

Figure 8-8
MW DEFERRED BY THERMAL STORAGE IN YEAR 2004 FOR LARGE OFFICES

Table 8-6 shows the results aggregated for the three North Carolina utilities. Without any utility programs, the market penetration of thermal storage in the year 2004 is estimated to be between 34 and 105 MW. With the alternative rate structures examined in this study, the market penetration estimates increase to 41 to 129 MW. With rebates of $200 per kW deferred, the estimates are 55 to 177 MW with partial storage systems, and potentially higher with demand-limited storage. The building types with the greatest potential are large retail, large offices, and secondary schools.

Table 8-6. ESTIMATES OF THERMAL STORAGE MARKET POTENTIAL IN NORTH CAROLINA

	MW Deferred in 2004	
	Low	High
Without any programs	34	105
With alternative rates	41	129
With $200/kW rebate	55	177

REFERENCES

[1]Battelle Columbus Division and Synergic Resources Corporation, Demand-Side Management, Vol. I - Overview of Key Issues, EA/EM-3597, Electric Power Research Institute, Palo Alto, CA.

[2]Limaye, Dilip R. and Todd D. Davis, "Implementation of Demand-Side Management Programs," IEEE Proceedings, 1985.

[3]Synergic Resources Corporation, Market Penetration of New End-Use Technologies, SRC Paper 906-166, Bala Cynwyd, PA. 1987.

[4]Synergic Resources Corporation, Analyzing Market Penetration of End-Use Technologies and Programs: A Review of Models, prepared for Electric Power Research Institute, May 1986.

[5]Synergic Resources Corporation, COMPEN Commercial Sector Market Penetration Model, SRC Paper 906-159, January 1987.

[6]Bonneville Power Administration and Synergic Resources Corporation, Industrial Electricity Conservation Potential in the Pacific Northwest, Final Report, Volume 1, March 1983.

[7]Zeidler, Peter C., "Estimating the Market Penetration of Cool Storage at Florida Power and Light," in Proceedings: International Load Management Conference, (prepared by Synergic Resources Corporation for Canadian Electrical Association and Electric Power Research Institute), June 1986.

[8]Wallin, Mark et al., Assessment of Cool Storage in Commercial and Institutional Buildings, Final Report prepared for North Carolina Alternative Energy Corporation, SRC Report No. 7272-R4, March 1987.

[9]McDonald, Craig et al., Commercial Cooling Market Study, Final Report prepared for San Diego Gas & Electric Company, 1987.

[10]Mahajan, Vijay and Robert A. Peterson, Models for Innovation Diffusion, Sage University Paper series on Quantitative Applications in the Social Sciences, series no. 07-048, Beverly Hills, California, and London: Sage Publications.

[11]Bass, F. "A New Product Growth Model for Consumer Durables," Management Science, January 1969.

[12]Mansfield, E. "Technical Change and the Rate of Imitation," Econometrica, 29 October 1961.

[13]Blackman, A. W., Jr., "The market dynamics of technological substitutions," Technological Forecasting and Social Change, 6: pp. 41-63, 1974.

[14]Fisher, J. C., and R. H. Pry, "A simple substitution model for chnological change," Technological Forecasting and Social Change, 2: pp. 75-88, 1971.

[15]Sharif, M. N., and C. Kabir, "A generalized model for forecasting technological substitution," Technological Forecasting and Social Change, 8: pp. 353-364, 1976.

[16]Rao, S., "An Empirical Comparison of Sales Forecasting Models," Journal of Product Innovation Management, December 1985.

[17]Lawrence, K. D., and W. H. Lawton, "Applications of Diffusion Models: Some Empirical Results," in New Product Forecasting, Y. Wind et al. (eds.), Lexington, Massachusetts: D. C. Heath, 1981, pp. 529-541.

[18]Synergic Resources Corporation, 1983 Survey of Utility End-Use Projects, Final Report prepared for Electric Power Research Institute, EPRI EM-3529, May 1984.

[19]Synergic Resources Corporation, Survey of Commercial Sector Demand-Side Management Activities, Final Report prepared for Electric Power Research Institute, EPRI EM-4142, July 1985.

[20]Synergic Resources Corporation, Survey of Utility Industrial Demand-Side Management Programs, Final Report prepared for Electric Power Research Institute, EPRI EM-4800, September 1986.

[21]Davis, Todd D. et al., Guidebook on Residential Customer Acceptance: Designing Programs That Attract Customers, Final Report prepared for Electric Power Research Institute (as a subcontractor to Battelle-Columbus Laboratory), SRC Report No. 7243-R3, December 1986.

CHAPTER 9
Competition

Clark W. Gellings
Louise M. Morman
Nancyann Emanuelson
Dennis Horgan

INTRODUCTION

It was coming -- some saw it, but we didn't listen. Competition -- real trench warfare[1] -- is now part of the electric utility jargon.

The electric utility industry is experiencing a major transformation. It is no longer safe to assume that electric utilities will continue to operate as they have, protected by regulators and secure in their service areas. Changes in regulatory policy and underlying economic shifts are spearheading the transition to an environment of increasing competition. This competition is coming from other utilities, from alternative energy suppliers, and -- indirectly -- from customers themselves.

Two key questions arise from a review of the competitive pressures facing electric utilities:

- What are the major forces of competition facing the electric utility industry today?

- What strategies are available to electric utilities operating in an increasingly competitive environment?

HISTORICAL FACTORS

Starting in the early 1900s, the industry entered a period of growth. Rural electrification, the Federal Power Act, and the creation of the federal systems (the Bonneville Power Administration and the Tennessee Valley Authority) set the stage for growth. The most significant legislation of that era, the Utility Holding Company Act of 1935, broke up large utility holding companies into regulated franchise operations and established an industry structure that has remained unchallenged for nearly 50 years.

The first two decades following World War II were a time of tremendous growth for the electric utility industry. The business environment was characterized by low annual inflation rates, low real interest rates, stable fuel costs, economies of scale in the construction and operation of electric power plants, and a regulatory climate friendly to public utilities. This business climate saw real electricity prices drop by 70% and electricity demand outpace economic growth. The major issues facing utilities were how to keep pace with rapidly increasing demand and how to incorporate advanced technological innovations in generation and transmission.

A series of events that began in the late 1960s radically changed the electric utilities' operating environment. Inflation accelerated, real interest rates peaked in the double-digits, and fuel costs rose to multiples of the pre-1970 levels,[2] primarily due to two major price shocks in 1973 and 1979. Gains in technological economies of scale diminished, and consumers became increasingly sensitive to environmental concerns. Significant legislation at the beginning of the period, including the National Environmental Policy Act (NEPA) and the Clean Air Act (CAA), encouraged the industry to spend billions of dollars on environmental controls, and affected the lead time and cost of utility construction programs. Near the end of the period, a variety of policy legislation, including the Public Utility Regulatory Policies Act (PURPA), the Powerplant and Industrial Fuel Use Act (PIFUA), the Energy Tax Act (ETA), and the National

Energy Conservation Policy Act (NECPA) of 1978, was enacted to promote alternate energy production and conservation efforts.

This legislation and the changing business climate rapidly cut demand growth for all customer classes, increased electricity rates from three to four times the 1970 average price per kWh,[3] and caused cutbacks in utility construction. The major issues facing utilities during this period were meeting environmental challenges, financing construction, and choosing fuels.

More recently, competitive market forces have become increasingly important. This business environment can best be characterized by reductions in inflation, rapid oil price decreases, significant advances in energy-saving technologies, a restructuring of the domestic economy, and a more aggressive regulatory climate.

Today's business climate also supports an increase in nonutility generation. There is a significant disparity in the financial positions found among utilities, with some being cash rich and others suffering financial problems due to continuing construction programs. Major issues facing electric utilities include:

- Preserving market position against gas utilities, oil suppliers, and other suppliers of generating services
- Effectively pricing services in this new environment
- Recovering costs of construction
- Participating in potentially profitable growth areas
- Successfully dealing with regulatory agencies.

Experience of Other Industries

Current trends in regulatory policy have set the stage for the growth of competitive market forces in the electric utility industry. As a means of promoting competition, the federal government is fostering policies which may remove limitations from public franchises and regulated monopolies and allow an increasing role for competitive market forces. The ultimate goal of open competition is lower prices for the consumer. Despite differences, analogies can be drawn to events in the airline, telecommunications, and financial services industries.

Fare and route deregulation in the airline industry forced airlines to compete on the basis of price as well as service. As a result, major industry restructuring continues to drive out uncompetitive carriers through mergers and bankruptcy. At the same time, new carriers continue to enter the market.

In the telecommunications industry, deregulation has led to the unbundling of telecommunications services and to major price reductions in long-distance service. Primarily through the breakup of AT&T in January 1984, long-distance service has been decoupled from local service, resulting in long-distance service price reductions. Examples of unbundled telecommunications services and prices include private ownership of customer-premises equipment and wiring, and separate access charges.

In the financial services industry, eliminating the ceiling on deposit interest rates has encouraged banks to compete for customer deposits on the basis of price rather than service. This major restructuring of the banking industry finds many banks attempting to improve profits by increasing market share. At the same time, banks are developing fee-producing services, such as credit cards and other ancillary services, that focus on customer needs. Additionally, nonbank institutions such as Sears Roebuck and Co., Merrill Lynch, and American Express are entering the banking market.

Seven clear trends -- with obvious implications for the utility industry -- can be gleaned from an examination of industries which have moved from a regulated to a more competitive environment:

- A market-oriented focus will be key to success.

- Profitability in the industry may decline as buyer power increases.

- More flexible pricing arrangements will evolve.

- New entrants will focus on market niches where high-value services can be provided at lower prices, threatening the profitability of entrenched competitors.

- Unaggressive companies will be left with expensive-to-serve customers and excess capacity, increasing the gap between winners and losers.

- Incumbents may be hindered in their ability to compete because they remain more regulated than new entrants.

- Merger and acquisition activity will increase as the industry restructures itself for competition.

Regulatory Policies and Economic Trends

The regulatory and economic trends that are removing limitations and promoting competition in other regulated industries are at work in the electric utility industry as well. For example, PURPA has led to Federal Energy Regulatory Commission (FERC) regulations which allow customers and third parties to cogenerate electricity and sell their excess at the utilities' cost. PIFUA has given cogenerators and industrial users preferred access to fuel. FERC Order 436 permits bypass of local gas distribution companies and direct sales to end users, increasing the attractiveness of natural gas as a substitute for electricity. Potential open bidding for service to federal facilities may mean both the loss of large customers and increased price competition.

Rising construction and fuel costs have led to higher electricity prices, increasing the attractiveness of conservation measures, energy efficiency, and alternative energy sources. Increasing cost differentials among utilities may cause industrial customers to relocate to gain cost advantages. Growth of the service sector of the economy, industry mobility, and excess generating capacity in some regions of the country have resulted in geographic competition for service to new industries and availability of low-marginal-cost power. Longer lead times for power plant construction have created financial problems for some utilities, greater difficulty in forecasting future power needs, and a trend toward modular generation.

These changes are not temporary phenomena. Utilities must position themselves, collectively and individually, to ensure that they can adapt to the new environment. Regulatory developments in the electric utility industry will have a substantial impact on the direction of competition. Utilities must remain close to their public policymakers, regulators, and legislators in order to have input into the development of policies responsive to utility and ratepayer interests.

THE SPECTRUM OF COMPETITION

The forces of competition are building in all areas of the electric utility marketplace. Discussing these forces in the context of a logical framework provides a mechanism for analyzing the nature and potential impact of each type of competition. Competitors can be grouped into three major categories which represent the spectrum of competition facing the electric utility industry:

- **Competition from other energy suppliers** -- Natural gas, oil, etc.; alternate energies

- **Competition from other electricity suppliers** -- Bypass, industry relocation

- **Competition from customers** -- Conservation, energy efficiency, cogeneration.

Other Energy Suppliers

This group of competitors represents nonelectricity energy suppliers, which can provide end users with the option of meeting their energy needs with sources other than electricity. Two major classifications of nonelectricity energy suppliers are natural gas, oil, and other traditional substitutes; and alternate energy sources.

Natural gas and oil are substitute energy sources which have traditionally competed with electricity. Natural gas distributors directly compete with electric utilities for space heating, water heating, and process heating markets, and are aggressively seeking to expand their market share through increased advertising and marketing activities. In addition, the Gas Research Institute and the Department of Energy are supporting research on thermally activated heat pumps which are expected to have coefficients of performance which will be double the efficiencies of the best current combustion systems. A supply surplus and regulatory changes supporting the bypass of local gas distribution companies are making natural gas increasingly price competitive. In certain regions, fuel oil has once again become a more attractive option due to recent oil price reductions. Figure 9-1 illustrates recent changes in efficiency for residential gas space heating systems.

Technical developments in solar heating, photovoltaic cells, biomass, wind power, geothermal power, and other alternate energy sources and systems are progressively reducing the cost and improving the efficiency of these alternatives; they are becoming more viable as long-term competitors with electric power.

Figure 9-1. AVERAGE EFFICIENCY RATES FOR GAS RESIDENTIAL SPACE HEATING SYSTEMS

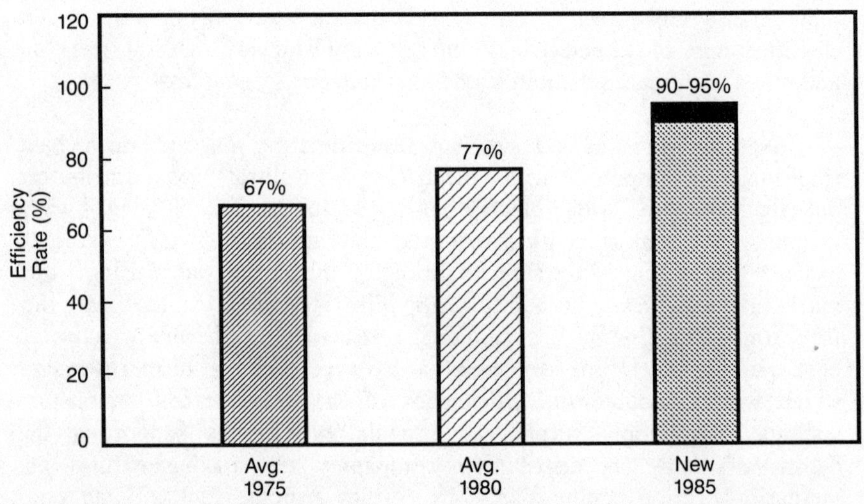

Source: Price Waterhouse[2]

Other Electricity Suppliers

The second broad category of competitors, other suppliers of electricity, affects demand for electricity in two ways: customers can bypass their own service territory's utilities to use power generated by other domestic utilities, independent producers, and foreign suppliers; and businesses can move out of a utility's service territory entirely in search of cheaper rates.

Competitive trends seem to indicate that "bypass" of traditional utility franchises may occur more frequently in the future. This is another example of the crumbling of utilities' historical boundaries.

Large industrial customers may be able to purchase less expensive power from out-of-region utilities offering lower rates. In fact, one Illinois municipality has found that it can purchase cheaper power across the Wisconsin state line. The Electricity Consumers Council (Elcon), the trade association of large industrial power users, has long argued that its members should be able to shop for power wherever the delivered cost is lowest; Elcon's efforts in this direction have been restrained by members' limited access to the transmission grid.[4]

The independent power market, comprising industrial and commercial producers as well as third-party cogenerators, has expanded rapidly in recent years. Such power has become competitively priced, and federal statistics, cited in The Wall Street Journal of February 16, 1986, indicate that independents may soon be generating 10% of the nation's power.[5]

Canada, possessing vast amounts of hydro power, offers some customers a low-cost alternative to their current utility's electricity supply. Mexico has also expressed interest in building plants to supply some U.S. needs.

Utilities have lost large customers due to the movement of businesses out of their service territories and the changing composition of the economy. Aggressive economic development programs are encouraging businesses to move their operations or to establish them in areas where they can realize economic benefits, such as discounted energy rates. Population shifts are occurring from the Northeast to the West and South. Plant closings and operating reductions are widespread in the manufacturing sector. The service sector of the economy is rapidly increasing its share of the gross national product, diminishing the share of energy-intensive manufacturing businesses. The transformation of Pittsburgh, for example, from a steel-producing economy to a largely service-oriented economy composed of banks, law firms, insurance companies, health care institutions, and other service-oriented businesses, has had a devastating effect on local electric utilities. America's manufacturing base has been eroded by foreign competition. When foreign facilities replace domestic production, domestic electricity sales are also replaced.

Customers

The final category of competition is customers themselves. Customers affect electricity demand through conservation and energy efficiency measures, particularly through cogeneration and other nonutility generation. Electric utilities often do not feel comfortable categorizing their customers as competitors. However, although not competitors in the traditional sense, customers are an important competitive force because their activities can reduce demand for electricity.

Conservation represents the willingness of customers to make either sacrifices or investments to reduce electricity consumption. Examples of conservation efforts include turning off lights and adjusting thermostat settings. In the 1970s, conservation was promoted as the long-term solution to the energy crisis. Today, conservation continues to influence the growth of demand for electricity. Demand is also affected by increased energy efficiency. New energy-efficient appliances allow customers to reduce the amount of electricity consumed without sacrificing the value (comfort or production) derived from those devices.

The concept has been around for decades, but cogeneration -- customers generating their own electricity -- became a significant competitive element only with the implementation of PURPA. Considerable growth in the number of cogeneration projects installed is projected for the next few decades, as illustrated in Figure 9-2.

Customers are not only producing their own power through cogeneration, but they have the potential to transport that power off-site through wheeling. Wheeling bypass -- the use of utility transmission facilities by third parties -- has profound implications for utilities. If transmission deregulation becomes a reality, third-party producers who cogenerate will be able to directly sell their power to utility customers, bypassing the utility except for a wheeling charge. Even without considering technical issues, this regulatory change would have dramatic effects on the level of competition by opening the grid to direct price competition.[6]

Figure 9-2. CAPACITY OF COGENERATION FACILITIES

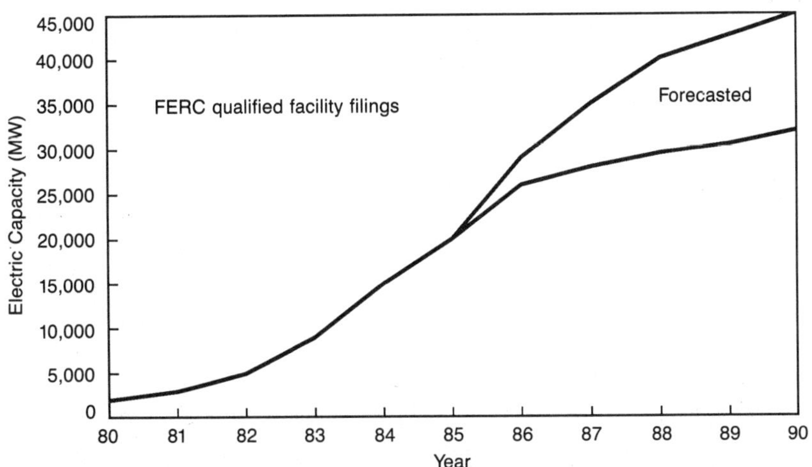

Source: Clark W. Gellings[6]

DEVELOPING COMPETITIVE STRATEGIES

The importance of competitors varies significantly by utility. Few utilities currently perform detailed competitor analyses, and many feel most uncomfortable about the effect of bypass, largely because it is new. Utilities need to examine all of the competitive forces affecting the industry and assess which require their greatest attention.

To successfully deal with the forces of competition and operate in this increasingly competitive environment, utilities must focus on their customers and become market-driven. To develop strategies with a market-driven focus, utilities must ask themselves two fundamental questions:

- Where are the best business opportunities?
- How do we position ourselves to compete?

Formulating effective answers to these questions will be the key to sustained competitive advantage in the 1980s and beyond. Figure 9-3 illustrates the basic decision-making process used to respond to these issues.

Figure 9-3. THE STRATEGIC DECISION-MAKING PROCESS

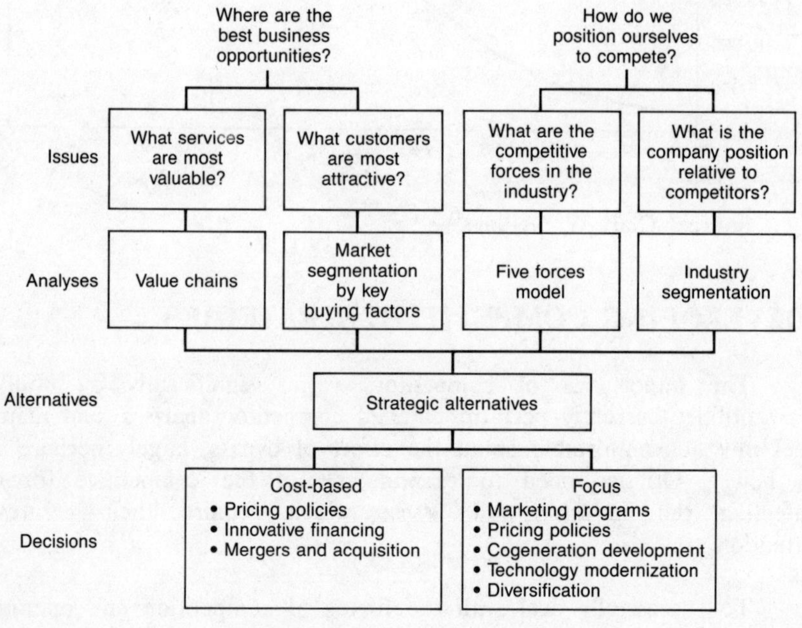

Source: Price Waterhouse[2]

Where Are the Best Business Opportunities?

To answer the question of where the best business opportunities lie, utilities must first identify what functions in their delivery of electricity to the customer provide the greatest value. For example, are generation, transmission, and distribution all sources of value to the utility, or is more value added by providing service, reliability, and equipment maintenance? Knowing where value is added in their delivery system will help utilities select the functional areas in which they wish to compete.

Utilities must then identify which customers are most attractive. It is increasingly important for utilities to understand their customers because of the lure of alternative sources of electricity (i.e., cogenerators, independent producers, and other electric utilities, both local and those in different parts of the country) and other energy sources (i.e., gas, oil, and alternate energies). Identifying the most attractive customers, capturing them, and keeping them are essential to profitability in the competitive arena.

How Do We Position Ourselves to Compete?

To answer this question, utilities must identify competitive forces in the industry and determine the extent to which those forces will affect them. This process should include identifying the key success factors which will determine the difference between winners and losers.

Utilities must also assess how well they are doing in relation to their competitors. The process of analyzing a company's position relative to competitors includes articulating the strengths and weaknesses of each competitor, and assessing methods for improving the company's competitive position. In conducting this analysis, utilities must consider that actions designed to enhance their position in an increasingly competitive environment may conflict with large social issues of interest to the regulatory community (e.g., providing high-value service to all customer classes). To maintain credibility with regulators, utilities must simultaneously pursue competitive efforts and public service responsibilities.

Evolving techniques of integrated value-based planning may reconcile these conflicting objectives. In value-based planning, the mix of supply-side and demand-side alternatives is balanced to meet society's energy needs at the maximum value.

Strategic Analysis Tools

Tools are available to help utilities perform the analyses described above, define available options, and make decisions regarding strategies to be implemented. These tools include:

- The value chain
- Market segmentation
- The "five forces model"
- Industry segmentation.

The Value Chain -- Traditionally, utilities provided generation, transmission, and distribution services that ended at the customer's meter. One of the effects of increased competition, however, is the breakdown of the functional boundaries of utilities. Laws or proposed regulations that mandate wheeling, require the purchase of independently produced power (PURPA), or allow preferential access to fuel (PIFUA) all act to break down the boundaries of traditional utility activities and allow entry of new competitors who may specialize in only one type of activity. In order to understand the implications of this type of competition, the utility needs to analyze the value of delivering electricity. The value chain is a mechanism for understanding and responding to the dynamics of an industry. Value chains divide a business or utility into functional activities to separate the cost and value of each activity.

In the electric utility industry, the traditional links in the value chain have been generation, transmission, and distribution. Electric utilities have historically performed these activities as an integrated service -- they purchased their fuel from suppliers; they purchased their technology and power plants from engineering contractors; and they transformed fuel into electricity and delivered it to the customer, directly or through distributors. Their relationship with the customer ended at the meter. Beyond the meter, the customer was responsible for satisfying his own end-use needs. In a noncompetitive environment, the traditional value chain was effective.

With the shift to a more competitive environment, however, the electric utility value chain is extending and its links are separating. In this new value chain, the functional activities of fuel supply, technology, generation, transmission, distribution, and customer end-use needs can be treated separately and then regrouped according to areas the utility identifies as providing its greatest sources of value.

As an illustration, consider a generation wholesaler who owns fuel and a power plant and is able to generate electricity; a transmission utility that transports bulk power for a fee; and a distribution utility that performs traditional distribution and service functions. All three could, in fact, sell end-use equipment on the customer's side of the meter. This hypothetical industry structure is significantly different from the traditional structure of the electric utility industry, but some elements of this value chain restructuring are already taking place. For example, according to an article in The Wall Street Journal of February 26, 1986, the San Diego Gas and Electric Company has vowed never to build another plant and will wheel power generated by Portland General Electric Company of Oregon and Public Service Company of New Mexico to satisfy distribution needs beyond its own generating capacity.[5]

An example of value chain fragmentation in another industry is the 1984 breakup of AT&T, which resulted in the separation of long-distance transmission of telephone signals from local transmission. Although some believe it unlikely that electric utilities will face pressures identical to those in the telecommunications industry, analogies can be drawn to proposed policies for mandatory wheeling and PURPA-mandated purchases of generated power.

The benefit of the value chain to strategic analysis is that it allows utilities to restructure themselves in ways that will provide greater value to customers and enable utilities to better adapt to the forces of competition.

Market Segmentation -- It is impossible to get close to customers if you don't know who they are and what they care about. Market segmentation divides the spectrum of customers into groupings of characteristics so that utilities can target customers, focus upon their individual needs, and prevent inroads by competitors. In this way, market segmentation forms the basis of an effective marketing effort.

Customers can be grouped by size or end-use characteristics. The most important grouping of customer classes in market segmentation, however, is key buying factors. Key buying factors are those characteristics that drive a customer to buy your product rather than a competitor's. This grouping categorizes customers according to purchasing behaviors.

Public attitudes toward the use of energy are not the same as they were before the energy dislocations of the 1970s. Historically, utilities have assumed reliability and cost of service to be the top priority for their customers. Research has shown, however, that customers are more interested in the value of electricity service for their specific end-use needs.[7] Thus, for the residential sector, major customer concerns are reflected in life-style services, such as comfort and convenience, in addition to cost minimization (Table 9-1).

In addition to grouping customers by buying factors, it is important to understand competitive influences by end use and to know how to counteract them. Some customers will always remain part of the core business; other customers will be subject to greater competitive pressures. For example, it is not sufficient to know that a particular customer segment is price-sensitive. It is essential to also know which competitor is likely to attempt to capture that business. In order to achieve this high level of customer knowledge, it is important to have focused information-gathering programs.

Table 9-1. CUSTOMER VALUES

Utility-Perceived Customer Values	Actual Customer Values
Reliability	Comfort
Cost	Convenience
Quality	Control
Power on demand	Appearance
Convenient billing	No hassle
Prompt service	Caring for me
Safety	Economy

Source: EPRI Journal[7]

As an illustration, one possible market segmentation technique for electric utilities would be to map residential, industrial, and commercial customers by their end uses, growth rates, buying factors, and sensitivity to price in order to determine the profitability and desirability of serving each customer group. Recent research suggests that mapping customers by functional preferences may be the best way to target end users.[7] These groupings are customer-oriented rather than end-use-oriented (the traditional approach). For example, grouping customers by preferences for functions such as comfort, economy, and reliability may be more logical than using more traditional rate classifications such as commercial and industrial.

The "Five Forces" Model -- To answer the question of who will satisfy customer energy-service needs, utilities must examine the competitive forces that affect the entire industry and determine their own position within the industry relative to their competitors.

Regulation, economic changes, changes in social and demographic characteristics, and developments in technology affect the entire industry. A tool for analyzing how these forces will affect competitors within the industry is the "five forces model" developed by Michael Porter of the Harvard Business School.[8] The forces of competition in Porter's model (as illustrated in Figure 9-4) are:

- Rivalry among competitors
- Bargaining power of suppliers
- Threat of substitutes
- Bargaining power of customers
- Threat of new entrants.

Rivalry Among Competitors -- This rivalry is determined by the nature of the product, the nature of the market, and the ways competitors interact with each other. For example, in industries with highly differentiated products (i.e., products significantly different from each other), the level of rivalry can be quite low. On the other hand, in industries with similar products and price competition, the level of rivalry is typically very high. Other factors that increase rivalry are high fixed costs within industries, high exit barriers to leaving an industry, and fragmentation of competition into numerous small competitors.

Based on these factors, one would expect rivalry in the electric utility industry to be intense. But regulation dividing service territories into franchises and controlling prices and returns has historically kept the level of cooperation among utilities high and the level of rivalry low. However, as regulatory controls weaken, making entry into the industry easier, more competition is likely to emerge among utilities.

Figure 9-4. THE "FIVE FORCES" MODEL

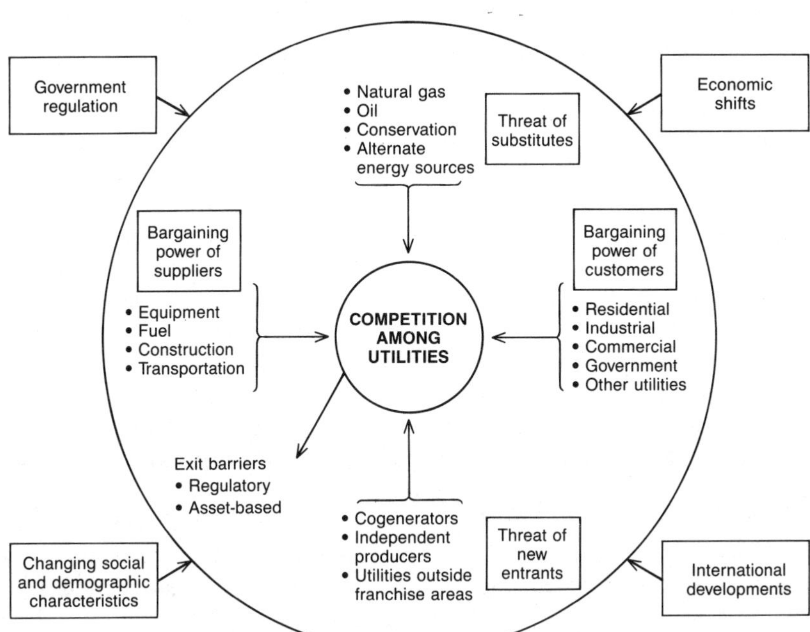

Who will benefit from changes in competitive forces is determined by the dynamics of the industry . . .

Source: Michael E. Porter[8]

Bargaining Power of Suppliers -- In some industries, suppliers have significant power. That power may come from their concentration, their ownership of proprietary products, or some other unique service. In the electric utility industry, supplier power has not been a major factor in determining the competitiveness of the industry. Some exceptions do exist, such as coal-fired utilities in the Northeast, for whom freight rates can be a significant cost factor. The competitive power of equipment, fuel, construction, and transportation suppliers has appeared to decline in recent years due to leveling off of fuel costs and cutbacks in new power plant construction. However, in a competitive market, there is an incentive for equipment suppliers and architect/engineering firms to forward-integrate by taking an equity position in cogeneration facilities or new central station plants.

Threat of Substitutes -- The profitability of electric suppliers depends on the availability and price of substitute energy sources to customers. Electric energy producers have traditionally faced competition from natural gas and oil; the level of that competition has varied as gas and oil prices changed. Recently, however, the natural gas industry has become much more aggressive in competing as a substitute for electricity, now that stable gas supplies seem assured. The prognosis for competition from alternate energy sources, such as solar power, is uncertain due to the end of tax credits and a decline in oil prices. In addition, the final threat of substitution is from customers who can reduce their levels of electricity consumption, substituting conservation for kilowatt hours.

Bargaining Power of Customers -- In the past, electric utility customers did not do much bargaining. Customers had few alternatives and, since the price of electricity was continuously declining, lacked strong incentive to investigate options. However, U.S. industry today faces intense, worldwide competition in steel, chemicals, and other basic product markets. Consequently, industry is attempting to obtain lower rates, gain access to lower-cost electricity sources, and substitute alternate fuels. With rising electricity prices, the availability of energy from cogenerators and independent producers, and the ability of customers to integrate backward and manufacture electricity, customers are in a much stronger bargaining position. Large customers have the option of relocating to areas with lower energy costs. Customers in some areas are also considering the formation of purchasing groups to increase their leverage with energy suppliers.

An example of the bargaining power customers can wield if they work together is the Citizen Power Group in Susanville, California. The group was formed to allow the community to find cheaper sources of electricity than the local power company, CP National Corporation.[5] Another illustration of how times have changed is the proposed policy calling for open bidding on utility service to federal facilities.

Threat of New Entrants -- Historically, regulation and technology virtually barred outsiders from entering the electricity industry. Utilities operated in regulated franchise areas, and economies of scale in generation presented a formidable entry barrier. Today, the combination of regulatory changes, such as PURPA, and developments in generating technology have made cogeneration and independent production more economical -- greatly lowering the barriers to entry and providing opportunities for increased competition.

Within the context of the five forces outlined above, important mobility barriers influence the ability of utilities to react to competition:

- Heavy capital intensity
- Switching costs
- Exit barriers.

Mobility barriers are important because unanticipated events such as abrupt regulatory action, loss of a dominant customer, or catastrophic equipment failure could allow a competitor to do substantial economic damage before the utility could effectively react. This possibility should convince utilities to prepare contingency plans that anticipate such eventualities, in order to gain as much lead time as possible. Electric utilities are capital-intensive and usually require long lead times to construct facilities. Thus, utilities change direction relatively slowly, while new entrants, such as cogenerators, can often move quickly to capture opportunity.

Once customers have invested in facilities to use electricity, gas, or oil, the high cost of changing these facilities over to use another form of energy influences the customer to retain his original energy choice. This incumbent advantage means that utilities that remain close to customers and carefully attend to their needs can expect to keep them from switching to substitute energy sources. A corollary strategy in the competitive world is to create switching costs by moving upstream into the customer's decision process and influencing the incorporation of electricity into new factories and homes. For example, Pennsylvania Power and Light Company encouraged a meat-packing company to purchase an electric smokehouse by sending an engineer to identify energy cost-saving ideas, and then arranging for a free vendor demonstration.[9] While switching costs are a barrier to substitution, they are not a barrier to other electricity suppliers that have access to customers through wheeling.

A third, and very significant, mobility barrier -- exit barriers -- impedes electric utilities which may wish to leave certain segments of the business. Regulators will not easily release utilities from their obligation to serve all customers within their franchise area, even if particular customer segments are not profitable. Thus, utilities must fight to keep their profitable customers and ensure a fair return from all customer groups.

Industry Segmentation -- The final analysis a utility must perform to formulate competitive strategies is an assessment of its position in relation to its competitors. Who are my competitors? What do they care about? Where are they going? What are they good at? How do I rank by comparison? These are questions utilities must answer before they can decide how to compete.

Industry segmentation, which is very similar to market segmentation, is a process that divides an industry into groupings defined by different operating attributes. These attributes vary widely according to utility and region. Some utilities are known for strong engineering and construction skills. Others have developed a tremendous base of low-incremental-generating-cost nuclear plants. Still others have thousands of miles of transmission lines, an obvious advantage in transporting power long distances.

In performing industry segmentation, utilities must assess their own key attributes and those of their competitors, such as:

- Cost position
- Financial resources
- Regulatory base and climate
- Customer base
- Reserve margin
- Technical skills
- Marketing strengths by customer segment.

Understanding these attributes facilitates the development of competitive alternatives.

Strategic Alternatives

After analyzing the cost and value of their activities, their customers' needs, the nature of competition in the industry, and their position in relation to competitors, utilities can begin to develop alternative strategies that will distinguish them in the marketplace. Two basic types of strategies apply to the electric utility industry:

- Cost-based strategies
- Focus strategies.

With cost-based strategies, a company competes by giving people more value for their dollar. With focus strategies, a company concentrates on getting closer to its customers, understanding and targeting their needs and providing distinctive services that keep competitors out and profits healthy. Utilities may choose to pursue a mixed strategy that employs cost-based elements in some operations and customer classes, and focus-based elements in others.

Cost-Based Strategies -- The production of electricity is a commodity business which has historically experienced significant marginal returns to scale. Consequently, the competitive energy marketplace tends to place a premium on cost competitiveness. A utility wants to start out as the lowest cost producer it can possibly be. The utility that can gain cost leadership will have a strong strategic position in the marketplace.

Economic turbulence in the electric utility industry in recent years has created significant differences in utilities' financial positions. Utilities that completed their construction projects and brought their power plants on-line without major cost overruns are now experiencing cash surpluses. These surpluses enable them to price aggressively and to pursue new business opportunities. Utilities still locked into expensive construction programs are net customers of cash and may have less flexibility. This financial disparity provides opportunities to develop both cost- and focus-based strategies that exploit these differences.

To develop cost-based strategies, utilities must understand their own costs and the costs facing their competitors. As they attempt to understand these costs, utilities should address the issues outlined in Table 9-2.

Cost-based strategies being implemented by utilities fall into two major categories: pricing strategies and capital restructuring. Many pricing strategies have been implemented in recent years. Incentive rates are now offered in 16 states. With discounts of up to 60% offered, these incentive programs are directed at new or expanding companies. Specific negotiated purchases and sales of power are becoming more common in the industry. Some utilities, faced with excess capacity, are adopting aggressive pricing positions to preserve their financial health.

Table 9-2
ISSUES IN DEVELOPING COST-BASED STRATEGIES

Customer Knowledge	Understanding Cost Drivers	Competitor Information	Process: Structuring to Minimize Costs	Linking Cost and Customer Focus	Management, Monitoring, and Control
Are segments appropriately defined?	What are the costs to serve different customers?	Who are competitors and potential entrants?	What are the most visible cost elements?	How should prices be set based on cost structure?	Does cost control reach top management attention?
How energy-intensive are the customers?	What are long-run and short-run marginal costs?	What are competitors' costs in long run and short run?	What choices exist for cost restructuring?	How do customers feed back information?	What key cost factors are monitored by management?
How price sensitive are the customers?	How do costs divide by functional area? By geographic area?	How do competitors' costs vary with factor prices?	What regulatory issues inhibit changes in cost structure?	Do actual profits by customer segment meet expectations?	How many layers of authority exist? What is the reporting structure?
What are customers' growth prospects?	What regulatory constraints are placed on cost allocation and recovery?	Do competitors have preferential access to any customer segment or distribution channel?	What systems are required to track costs?	How quickly can prices be changed?	How are the competition's actions monitored?
What reliability does the customer want?	How sensitive are costs to factor prices (e.g., fuel, capital)?			Who has pricing authority and profit responsibility?	
What additional services would be attractive to the customer?					

These pricing strategies are being applied, for the most part, to the industrial sector. Pacific Gas and Electric Company (PGandE), for example, recently announced rate cuts of 23% for large business customers versus 7% for residential customers.[10] Similarly, Iowa Public Service Company recently raised residential rates nearly 15%, while lowering some commercial rates more than 19%.[10] Commonwealth Edison Company has offered discount rates for electricity to its 100 largest industrial customers in excess of their average consumption of the last three years.[10] Detroit Edison Company has designed a metal-melting rate to promote electricity sales to steel producers.[9] These pricing policies, while bringing prices more in line with the marginal costs of providing service to different customer classes, are directed at preserving or increasing the customer base and preserving revenue that might otherwise be lost.

The second major category of cost-based strategy is capital restructuring. Corporate reorganization through sale and leaseback of capacity, and spin-off or sale of packaged capacity, support the separation of generation from transmission and distribution (i.e., fragmentation of the value chain). Mergers and acquisitions facilitate reductions in overall utility costs.

One example of capital restructuring in the industry is the sale by the Deseret Generation and Transmission Cooperative of its Bonanza Power Plant to a unit of Shell Oil Company, and Deseret's subsequent leaseback of the plant. Deseret estimates that the arrangement will reduce its annual costs by as much as $50 million.[11] Portland General Electric Company sold its stakes in a coal plant and power line to General Electric Company's credit subsidiary, and then leased them back. That arrangement is expected to help the utility offset projected cost increases.[12] Nuclear plants in New England and Kansas have been organized as separate companies so that they can realize the benefits of scale and professional management that large nuclear utilities, such as Duke Power Company, have shown are possible in multiplant operations.[9] The merger of Cleveland Electric Illuminating and Toledo Edison into Centerios is forecasted to save as much as $1.3 billion through reduced overhead and more efficient power plant operation.[9]

Focus Strategies -- To survive and be successful in a competitive environment, utilities must focus on their market and devise strategies for identifying, capturing, and keeping desired customers. Focus strategies include targeting attractive customer segments, creating customer switching costs, and diversifying utility operations.

Targeting customer segments helps the utility become market-driven. To become market-driven, utilities must segregate their customers and provide them with much more than just electricity delivery at the meter. Traditional marketing functions must be revived, customer account teams established, customer industry expertise developed, and services offered beyond the meter. Segment focus can be combined with custom pricing for troubled industries or for other specialized customer needs. Focus can also enable a utility to differentiate itself from competitors by providing customized services, such as energy management consulting and equipment and systems design.

Focus can also mean concentrating on specific business segments within a service territory. For example, utilities may choose to concentrate on generation and then transmit wholesale power to others. Alternatively, utilities may concentrate on distribution to customers and develop a superior network to efficiently service those customers.

PGandE is one example of a utility that has implemented focus strategies to target customer segments. The company has decentralized its operations and reorganized them along customer lines. A major accounts program has been created to provide individualized attention to PGandE's biggest customers. One outgrowth of that effort is a pilot program to help large customers tailor their electricity usage to a price schedule that fluctuates by time of day. When it cannot prevent a customer's move to self-generation, PGandE will consider whether to participate in developing the cogenerator's plant.

In order to create switching costs, utilities must be in a position to influence customers' long-term energy decisions and manage prices to prevent entry by competitors. As previously discussed, once customers have invested in facilities for a certain type of energy, the cost of switching facilities to accommodate another form of energy influences the customer to retain his original energy choice. Thus, it is important for utilities to be involved in their customers' initial decision-making process. Additionally, if the utility can keep its prices competitive with other energy sources, the customer will have little incentive to switch.[13]

Technology modernization is a means of influencing customers to choose and retain electricity. By financing the development and refinement of end-use electrical equipment and processes, utilities can encourage customers to use electricity in areas where they may not have previously considered electricity. Examples of technology modernization efforts by electric utilities are the Center for Metal Fabrication at the Battelle Memorial Institute in Columbus, Ohio, and the Center for Materials Production at the Mellon Institute in Pittsburgh, Pennsylvania. The centers, funded by the Electric Power Research Institute (EPRI), were created to encourage the use of energy-efficient, electric-based processes by metal fabricators and producers.

The cash surpluses of utilities and, to some extent, competitive trends in the electric industry are leading to diversification. For example, Florida Power and Light Company has acquired a life insurance company, Colonial Penn Group.[5] Pacific Lighting Corporation has acquired Thrifty Corporation, a drugstore chain.[14] San Diego Gas and Electric Company is selling computer software used in its utility operations and investing in alternative energy projects and real estate.[14]

To develop focus strategies, utilities must understand their customers' needs. As they attempt to understand these needs, utilities should address the issues outlined in Table 9-3.

FUTURE RESEARCH

As utilities adapt to a more competitive environment, the focus of their activities will change. Utilities will place increasing emphasis on cost analysis and control; they will become more customer-focused; and they will become increasingly driven by the search for profit-making opportunities. As utilities change direction, the focus of research must also change.

General Considerations

As the utility industry becomes more competitive, several considerations will become increasingly important to utility executives as they evaluate their research and development investments:

Figure 9-5 illustrates a possible taxonomy of research options for the industry in its transition to a more competitive environment.

- **Technology as a competitive weapon** -- Utility executives will be looking for products that will distinguish them from or lower their costs in relation to their direct competitors, and substitutes like natural gas. These technologies include demand-side technology programs, such as on-site thermal storage or distributed generation technology, and supply-side technologies, such as advanced fluidized bed coal combustion.

- **Research to improve customer focus** -- In a competitive world, research that improves a utility's knowledge of its customers and permits better customer targeting will become more important. Examples include research in the "soft" sciences, such as demand-side planning, forecasting, and market analysis.

Table 9-3. ISSUES IN DEVELOPING FOCUS STRATEGIES

Customer Knowledge	Process: Meeting Customer Needs	Resources: People and Tools	Defending Against Competitors	Execution	Management, Monitoring, and Control
Who is the decision-maker?	What are the key tasks in the sales cycle? Who is responsible for each?	What are the key skills necessary in sales?	How are leads matched to the appropriate sales people?	How is performance measured?	What key factors are monitored by management?
How are customers prioritized?	What is the length of the sales cycle? What are the principal components?	What are the characteristics of the sales force (e.g., experience, age, turnover, number of accounts)?	How is intelligence shared through the sales force? What research is performed?	Are actual revenues equal to potential/ expected revenues?	How many levels of authority exist? What is the process for approval?
What are customers' key buying factors?	How is the sales force structured? Does this follow the way the market is segmented?	What is the profile for a top performer (e.g., more time at customer site, more experience, better use of sales tools)?	Which sales tools are used for which accounts?	What is done for customer follow-up?	What is the reporting structure? Who bears direct responsibility for customer satisfaction?
Are segments appropriately defined?	How are leads generated? How is market intelligence collected and distributed?	What tools are used to generate sales?	How are resources budgeted/al-located?	How do customers feed back information (e.g., surveys or questionnaires)? How is this information used?	How are accounts managed? Who decides how an account will be managed and coordinates deployment of resources?
Which customers are more profitable than others?	How are leads evaluated, ordered, assigned?	What sales support is offered to the sales force?	How is the determination made to include higher levels of management in negotiations?		
			How are sales linked to planning, order entry, technical support, service?		

Figure 9-5
TAXONOMY OF RESEARCH OPTIONS

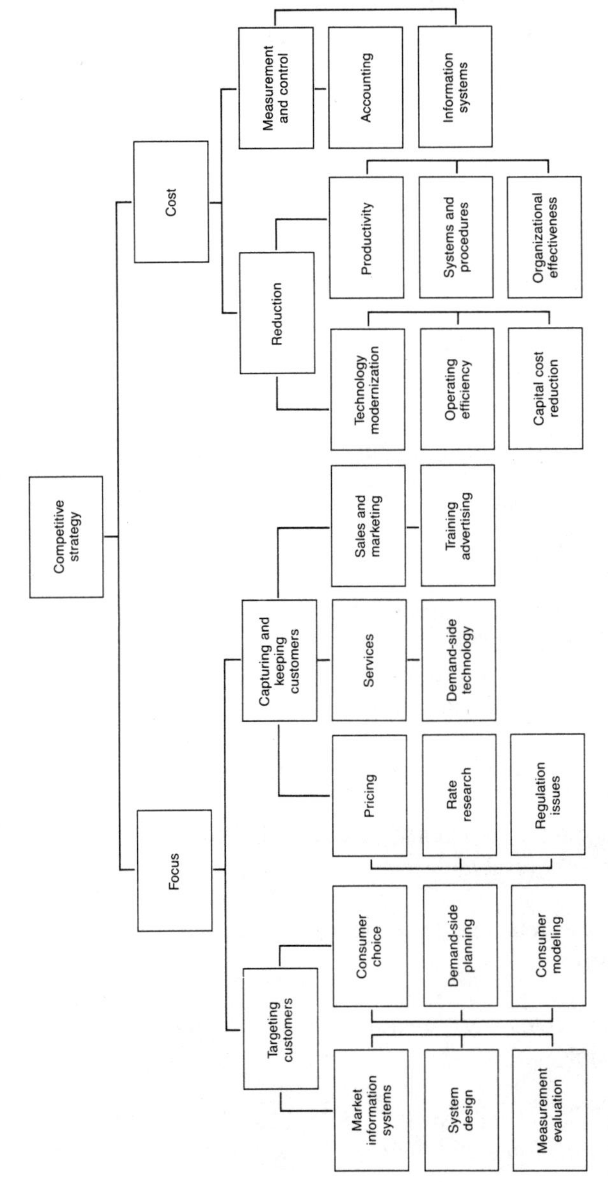

Source: Price Waterhouse[2]

- **Research to improve productivity** -- Research to improve the productivity of a utility's assets and its labor force -- thereby increasing profitability and improving competitive position by reducing costs -- will become more urgent. For example, increasing emphasis may be placed on programs to extend generator life, increase the capacity of existing transmission, and automate customer services to reduce delivery cost.

These three considerations will necessitate the kind of studies that have not traditionally been a major part of utility research programs (e.g., research in management science, accounting, and behavioral sciences). It is important to consider that in the competitive world of the future, research programs will be driven not only by technology requirements, but by business policy requirements as well.

CONCLUSIONS

The following conclusions summarize this chapter on competition in the electric utility industry:

- There is no single and certain answer to competition.

- Competitive forces are real and are here to stay.

- Substantial evidence exists of trends toward competition across several industries.

- Public reaction to economic deregulation has generally been positive.

- The rules of operation are changing; utilities must be flexible.

- Utilities need to get closer to their customers and gather information about their buying patterns and intentions.

- Fellow utilities today may be competitors tomorrow; thus, utilities need to learn all they can about one another.

- Utilities will need to restructure their organizations and develop cost information systems to survive in a less regulated environment.

- In this period of competitive transition, changes may occur faster than utilities can react. Utilities must stay close to the regulatory process to ensure they are not "blind-sided" by arbitrary action.

ACKNOWLEDGEMENT

This chapter is based in part on a report prepared by Price Waterhouse for the Electric Power Research Institute under research project RP 2381-6 and published under the title "Competition: Pressures for Change," EPRI report number EM-5226, May 1987.

REFERENCES

[1]Richard McCormack, "It's Trench Warfare for Gas and Power," The Energy Daily, January 6, 1987.

[2]Price Waterhouse, Competition: Pressures for Change, EM-5226, Electric Power Research Institute, Palo Alto, CA, May 1987.

[3]Arthur A. Thompson, Jr., "New Driving Forces in the Electric Energy Marketplace -- To A 'Death Spiral" or Vigorous Competition?" Public Utilities Fortnightly, June 21, 1984, pp. 31-40.

[4]"Competition," TEL, The Electric Letter, April 7, 1986.

[5]Bill Paul, "Power Play: Electric Utilities Find Market Forces Taking More Important Role," The Wall Street Journal, February 26, 1986, p.1.

[6]Clark W. Gellings, "Cogeneration & Electric Utilities: Current Status, Regulatory & Competition Issues," Electric Power Research Institute, September 1986.

[7]"Understanding the Consumer," EPRI Journal, October 1986.

[8]Michael E. Porter, Competitive Strategy, The Free Press, 1980.

[9]John C. Sawhill and Lester P. Silverman, "Your Local Utility Will Never Be The Same," The Wall Street Journal, January 2, 1986, p. 10.

[10]Bill Paul, "Electric Rates for Industry Are Being Cut; But Homeowners Are Paying a Greater Share," The Wall Street Journal, April 4, 1986.

[11]"Shell Unit Buys Power Facility for $664 Million," The Wall Street Journal, January 3, 1986, p.2.

[12]"Portland General Sells to GE Stakes in Plant and Line," The Wall Street Journal, January 3, 1986, p. 28.

[13]John W. Wilson and Barbara Starr, "Pacific Gas & Electric's Crash Course in Competition," Business Week, March 3, 1986, pp. 116-117.

[14]Frederick Rose, "Utilities, Flush with Cash, Enter New Fields; But Some Fear Diversification Push Might Go Too Far," The Wall Street Journal, July 1, 1986, p. 6.

CHAPTER 10

Converting From Conservation Programs to a Marketing Mentality: Can the Two Coexist?

John H. Chamberlin and Nancy Licht

INTRODUCTION

During the 1970s, there was a marked surge in the development of conservation programs offered by utilities in the U.S. After decades of promotional efforts, utilities responded to the increases in fuel prices, capital costs, and rates which marked the early 1970s by developing a complex and comprehensive set of programs designed to encourage customers to adopt technologies, measures, and practices which would reduce their use of electricity.

Some of the impetus for the development of these programs was political. Conservation programs lessened the impact of the bill increases that faced most customers. Many of these programs, however, were offered for cost-effectiveness reasons. The costs of the programs, less the reduction in revenues to the utility, were less than the costs of electricity which would otherwise have been provided.

Whatever their reasons, most utilities in the U.S. developed a significant ability to plan, evaluate, and implement conservation programs. Staffs grew, and organizations fostering conservation efforts were developed. However, as the 1980s began, three significant factors began to emerge which altered this emphasis on conservation. First, many of the lower cost conservation actions (such as installation of water heater blankets and thermostat setbacks) had already occurred. Further actions would entail higher per unit costs. Second, and more importantly, the cost structure of many utilities began to change. Gas and oil prices declined significantly. Indeed, costs of most fuels fell significantly between 1982 and 1986. These reduced costs are projected to remain in place for the foreseeable future. The capacity additions which took place during the 1970s, combined with slower than anticipated increases in sales, have left many utilities in a position of excess capacity. Conservation programs, as a result, are often not cost effective. However, increases in sales (i.e., marketing programs) produce greater revenue than the sales efforts entail in generation costs. Finally, utilities increasingly recognize the competitive environment in which they operate. Larger, more profitable customers threaten (and follow through) with relocation decisions, or actions to become self-sufficient in energy production. Loss of these customers means reductions in net earnings.

Utilities have responded to these changes by developing marketing programs designed to identify and penetrate profitable sales areas. At the same time, these utilities also maintain conservation programs, staff, and organizations. In many cases, these organizations and programs work at cross-purposes. The purpose of this chapter is to identify ways in which conservation programs can remain attractive and profitable entities within a utility focused upon marketing efforts. To what extent can these conservation programs assist in the achievement of marketing goals?

First, we will review the rationales for marketing and conservation programs. Then, the role of existing conservation programs within a marketing environment will be discussed. Organizational changes required to make a successful transition from conservation to marketing will be outlined. Methods available to modify, refine, or retain conservation programs will be described. Finally, the particular role of rate designs will be addressed.

The entire discussion is directed toward the goal of building the capability to accomplish marketing goals in the broadest sense. While the specific load shape objectives of an organization will necessarily change over time (as market conditions change), the broader goal must always be to implement effective programs which influence customers to take desired actions. Whether these are conservation or marketing actions depends upon the specific situation.

SELECTING LOAD SHAPE OBJECTIVES THAT ACHIEVE CORPORATE GOALS

Whether any utility should pursue conservation or marketing programs depends solely upon the broader goals which are sought (Figure 10-1). Neither type of program is intrinsically desirable. Conservation programs are not desirable simply because they reduce the energy consumption of customers. Rather, depending upon the perspective of the observer, they are desirable because they reduce the average level of rates, reduce the need for new capacity, reduce the amount of oil imported into the U.S., or reduce the amount of pollutants in the environment. Marketing programs can be evaluated upon the same basis.

In general, cost-effectiveness analysis is concerned with assessing the extent to which either conservation or marketing programs are desirable. A widely (though not unanimously accepted) standard is that programs are "desirable" if they pass the "impact-upon-rate-level" test. That is, a conservation program is beneficial if it reduces the average level of rates below what it would otherwise be.

In practice, the test requires that the cost of the program, minus the revenue loss associated with the change in consumption, be less than the cost of the electricity which would have been otherwise generated. This test is usually performed by comparing the program cost and revenue loss with the marginal costs of generation.

Marketing programs are evaluated similarly. The benefit of the program is the increase in revenue associated with expanded sales. The costs include the cost of the program itself as well as the marginal costs of the increased generation. In both cases, marginal costs include the marginal costs of capacity as well as energy.

Figure 10-1. LOAD SHAPE OBJECTIVES DEPEND ON MARKET CONDITIONS

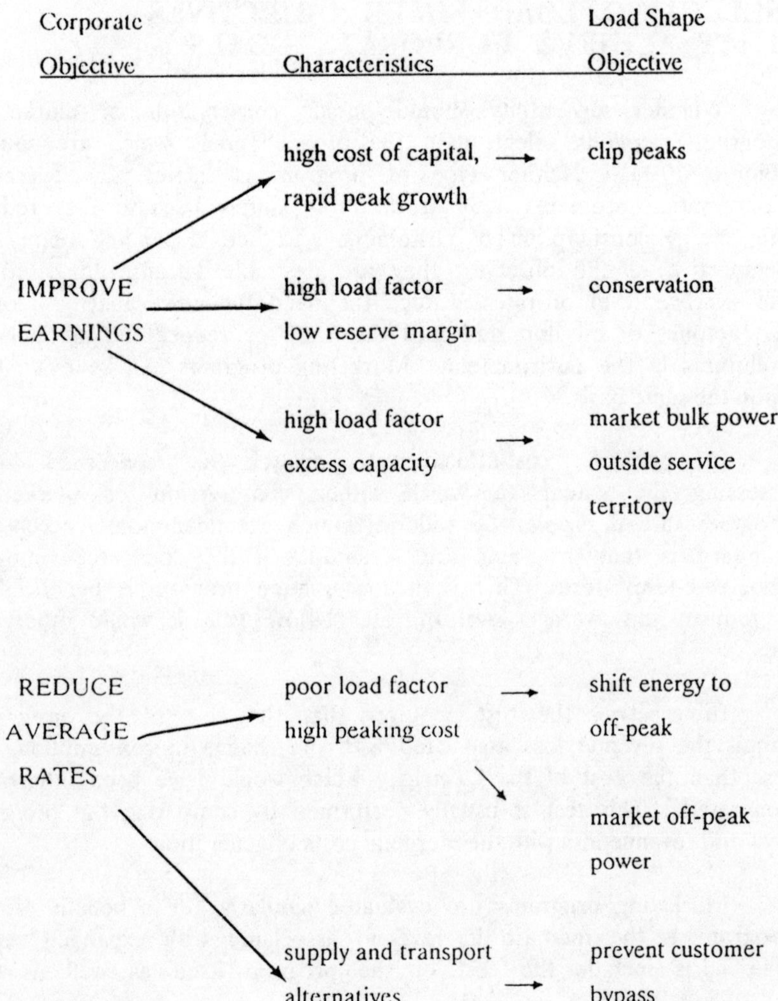

Thus, the decision about whether conservation or marketing programs should be pursued can be made in terms of a cost-effectiveness standard. However, this cost-effectiveness perspective can easily be implemented too narrowly; both types of programs have a significantly more complex relationship to the achievement of broad corporate goals.

CAN CONSERVATION PROGRAMS BE ATTRACTIVE IN A MARKETING ENVIRONMENT?

Even when a utility has determined that conservation programs generally do not achieve cost-effectiveness goals, there may still be a significant role for these programs. This chapter discusses the ways in which conservation programs may provide benefits to the utility with a marketing focus.

Variation in Load Patterns -- Even when marketing is cost-effective, increases in all loads are not generally cost-effective. For example, water heater loads typically peak during morning hours, and air conditioning loads during the afternoon or early evening. A utility may have variations in marginal costs such that increases in water heater loads are profitable, while increases in air conditioning loads are not. In such a case, home insulation programs may be cost-effective at the same time that programs to increase the adoption and use of electric water heaters are profitable.

Even in a case in which the average diversified load for a particular end use is profitable, specific segments of the customer population may have unprofitable loads. Air conditioning use for a particular upwardly mobile type of customer may not be profitable, while air conditioning in general may be. These differences arise because of different use patterns among segments of the population.

Thus, it is critical to identify the load patterns of specific end uses and population segments. For some of these segments, marketing programs may be cost-effective. For others, conservation programs will be appropriate. This is only inconsistent if each program is viewed as being valuable for its own purpose. As long as the overall goal is to improve the profitability of the company, or to reduce the average level of rates, different types of programs will be appropriate in different circumstances.

Timing Variations -- Programs will have different values across time. For example, a marketing program will generally give rise to both long-term energy and capacity requirements. Many utilities are currently in capacity surplus situations, which may continue for the next half decade. By the mid-1990s, many utilities will again be in capacity deficiency positions. As a result, the impact upon a future need to build capacity resulting from marketing programs must be considered. Thus, programs which forestall this long-term need have value. Programs which effectively sell energy while conserving capacity can coexist with marketing efforts. For example, interruptible rates establish a mechanism to reduce coincident loads of customers, while promoting usage through discounted bills. Also, efforts to increase the installation of energy management equipment may allow future participation in interruptible programs.

Insurance -- Some utilities view conservation programs as insurance. These programs take many years to put into place. Customers must be recruited, billing systems established, utility representatives hired, participants' attitudes changed, equipment tested, purchased, and installed, etc. If load were to grow at faster rates than anticipated, or if fuel prices were to rise rapidly again, utilities which have profitable marketing programs may once again be strongly interested in conservation. Maintaining an active conservation effort, even when it is not currently cost-effective, provides insurance against being caught unprepared by sudden shifts in the market.

Low Income Assistance -- In many states, the requirement for utilities to provide assistance to low income customers has remained in place despite changing market conditions. When properly segmented, conservation programs can provide this assistance while improving the company's financial position. To the extent that a utility has flexibility regarding participation in low income assistance programs, conservation programs can be tailored to reduce such customers' least profitable loads. For example, it may be far more cost-effective to reduce air conditioning usage than space heating usage.

Interfuel Competition -- Many utilities provide both gas and electricity. Often, the gas territory and the electric territory do not completely coincide. In such cases, it often makes sense to take different postures toward conservation in the overlapping parts of the service area. For example, a weatherization program in areas where only gas is served might do less harm to the utility than would the identical program in overlapping areas. This is true because the program would reduce the sales of only one of the products, rather than both.

Impact Upon the Customer -- Conservation programs are now important elements of many utility product offerings. Some customers achieve a strong sense of satisfaction through having the opportunity to conserve. People also obviously like the option of having something they can do to control their energy bills. Conservation programs thus have a significant effect on public and customer relations, an effect which should not be ignored. Again, this argues for understanding the specific load impacts of conservation programs, and for having the ability to design programs that will allow conservation to occur at the least cost to the utility.

MAKING THE TRANSITION FROM CONSERVATION TO STRATEGIC MARKETING

While the organizational and operating structures found in most utilities today can support a new marketing mentality, there are a number of issues which must be resolved to achieve a smooth transition. These obstacles reflect the market environment in which conservation activities were created and the manner in which they have been instituted in most organizations. This section describes several important implementation barriers, and outlines organizational and programmatic issues involved in the conversion from a conservation orientation to a marketing orientation.

Problems of Integration and Influence

The conversion to a marketing mentality is hindered by both conceptual and operational problems. Most conservation programs, developed in the crisis mode of the mid-70s, are not well integrated into utility operations, but instead exist in separate parts of the company. Residential programs are often in one department; commercial/industrial programs in another. Load management activities may be in the rate department while rebates and education are in customer operations or service units.

As a result, there are often no program development, data collection, or evaluation functions that span all customer classes or even address all programs and strategies for a single class. This has clearly increased operating costs through duplication of effort. More importantly, it has impaired the utility's ability to accurately assess the total impact of conservation and fully incorporate conservation results into resource plans.

A report prepared for the Edison Electric Institute's (EEI) National Marketing Program[1] noted that marketing as a concept and organizational entity was literally expunged from all corporate records in the early 1970s and that marketing personnel were scattered throughout the company. In many cases, these actions were encouraged by regulators and supported by ratepayers. As a result, top utility management may be hesitant to once again embrace activities that appear marketing oriented.

Georgia Power Company executives interviewed for the EEI study point to two additional implications of the deemphasis of marketing. First, the successor activities to marketing were unable to analyze and predict major market trends, and as a result assumed a reactive mode. And second, rate increases became the principal means of dealing with revenue shortfalls.

In many cases, conservation entities have acted as advocates within the company and have based their actions on a specific philosophy and view of the marketplace. Now, a significantly different philosophy is being superimposed on these groups.

Finally, most utility conservation programs have addressed almost all customer classes and end uses. Today's marketing efforts are more selective and complex, with activities focusing on retention of large commercial and industrial customers with alternative supply options, and building specific kinds of loads. Moreover, many commercial and industrial customers require a significantly broader set of services than in the past. Rate analysis and negotiation, energy management assistance, economic development, wheeling, transmission, and bulk purchase issues all fall under the purview of the new marketing representative. These factors affect the type of data that should be collected, the staff expertise required, and the overall position of marketing within the corporation.

Organizational Changes

The successful conversion to a marketing orientation depends upon an understanding of the evolving utility environment and the design of an organizational structure that keeps pace with market changes. To ensure its success, the new marketing-oriented structure should be well-integrated with overall corporate objectives and visibly supported by top management. Utilities also should develop clear systems to communicate the benefits of marketing, both internally and externally, and should review and possibly redirect their field forces.

Sales programs and demand-side programs are each tools which, depending on market conditions, can be used to achieve corporate profitability and customer satisfaction. In order to select the most appropriate mix of and timing for these alternative programs, marketing managers must receive direction and information from supply, finance, rates, regulatory, and other utility departments, Moreover, effective implementation of the most beneficial set of activities involves many areas of utility operations including corporate planning, communications (internal and external), research and evaluation, customer service, field operations, and rate design. One method of achieving the necessary integration is to place marketing at a high level within the organization.

Developing a strong, centralized marketing organization is critical. A significant lesson learned from the development of conservation programs is that fragmented, isolated programs lead to difficulties in planning, selection, and implementation of programs. There needs to be significant coordination of rates, load research, market research, planning, evaluation, and implementation functions. This can only occur under a centralized organization.

For example, it is critical to the selection of programs that useful market research information be developed. Understanding what kinds of customers are willing to participate in rebate programs, for example, and what kinds are not allows market segmentation which can lead to improved participation rates, more cost effective promotional efforts, and greater customer satisfaction. Yet market research groups may not provide the kind of information planners seek, particularly in a fragmented organization. They may resist changing the questions on the customer survey, perhaps because of a desire to maintain consistency from year to year, or to allow analysis of issues of concern to some other part of the company. As a result, marketing planners may not have the information they need to design the most effective programs.

Utilities also need to communicate marketing objectives and benefits to three major audiences: employees, customers, and trade allies. Marketing itself to its employees is a difficult task for a utility for two reasons. First, conservation employees often have strong beliefs that conservation programs are intrinsically good. These beliefs were developed during a time when the cost-effectiveness rationale of these programs was so strong that it became "assumed"; it was forgotten that the underlying conditions could change. Second, conservation departments have generally remained relatively isolated from the mainstream of the utility organization. Often, the conservation groups had to struggle to have their programs adopted and to obtain the requisite funding. As a result, conservation employees are often strong advocates for their programs.

These factors can make the transition from a conservation program to a marketing program particulary difficult. Employees will find it difficult to accept. Therefore, an aggressive internal campaign is required to ensure that all employees understand the logic behind the transition. This may require training in cost-effectiveness analysis as well as reviews of changes in market conditions.

Customer awareness is also a critical factor in the transition from conservation to marketing. For example, Lincoln Electric System (LES) in Nebraska initiated an extensive advertising and public relations campaign in advance of the implementation of a marketing program. The utility's goal was to ensure that customers understood that a primary goal of its load shaping program was reducing future rate increases. LES urged its customers to increase the wise and efficient use of off-peak electricity and communicated that its marketing program was complementary with its existing conservation efforts. It is critical to ensure that customers understand why the same utility which paid them to insulate their homes two years ago is now encouraging them to increase their use.

Another example: Puget Sound Power and Light's Customer Council, established in 1980 to advise on conservation matters, increasingly addresses marketing issues, such as impending competition in the Northwest natural gas market.

Trade allies represent another key audience for marketing messages, particularly given the fact that many new marketing efforts will be directed at the commercial and industrial sectors. Utilities should reinforce their working relationships with building and appliance industry representatives.

Customer contact through field offices is one of the most important aspects of marketing implementation. As utilities begin new marketing efforts, field staff should be trained in conventional sales techniques. Additionally, a strategic accounts system can ensure personalized contact with major customers.

Pacific Gas and Electric Company (PGandE) established an account executive program to provide a single point of contact for the utility's largest commercial and industrial customers Each of the 12 accounts managers specializes in one industry area, and provides clients with services ranging from energy management to rate negotiation. These managers are responsible for tracking and reporting competitive threats and industry trends. PGandE top management expects its field managers to personally know the major customers in their regions and divisions, what PGandE's competitive advantages are, and what the competitors have to offer.

At Pacific Power and Light Company in Oregon, the utility's senior managers, including the president, each received responsibility for several strategic accounts. These managers call on customers, participate in negotiations, and report on account activity. Pacific Power conducted a series of sales training sessions to teach managers about sales plans, account classification, and other traditional marketing techniques.

Load evaluations and opinion research are increasingly important due to the need to selectively market the utility's product. Developing a system for gathering competitive intelligence will further aid in identifying which marketing strategies to develop for particular customers.

WHY LOAD EVALUATION IS IMPORTANT

Many utilities have developed significant capabilities for estimating the load impacts of conservation programs -- capabilities which can also be used to design the best possible marketing effort.

A primary reason for understanding the effect of load impacts on either a conservation or marketing program at a reasonably detailed level is to facilitate program design. Regardless of the type of program offered, it is difficult to design it in the most cost-effective way without understanding its load impact. The information gained from load evaluation is essential to the design of a program that will get the most for the program dollars. This is true for the design of new programs as well as the redesign or redirection of an existing program.

A program's cost-effectiveness depends upon whether the benefits exceed the costs. The load impacts of a program are very important factors in determining the cost-effectiveness of any type of program. The increased cost or savings in the production of energy and demand is the product of the marginal costs (of generation, capacity, and transmission and distribution) and the load impacts. Since these marginal costs can change dramatically by season and by time of day, the timing as well as the size of the impacts is important. The revenue change, which is the benefit of a marketing program and a cost of a conservation or load management program, is comprised of the rates times the load impacts. The last component, the cost of implementing and operating a program, is reasonably predictable, and is rarely as large as the total cost of producing the energy and demand sales or the revenue change. Thus, the load impacts from a program are a significant element in determining whether a program is cost-effective, as well as <u>how</u> beneficial it will be.

A utility can discover a lot of information about its customers' loads by performing a load study. But only an evaluation of the load impacts of various types of load modification programs can tell how these loads are affected when purposely influenced, such as through a conservation or marketing program.

Since energy is actually consumed by end-use equipment or appliances, rather than by the customers, all demand-side programs depend on a change in either the quantity or size of these appliances or in the pattern of use of the appliances. This is true whether the program is intended to decrease energy use, to clip and shift peak loads, or to increase energy use. A program may be designed to cause a certain change in the types of appliances used or in the patterns of appliance use, but whether it achieves this goal is uncertain. This uncertainty exists for two main reasons: not knowing what appliances actually exist in the market and what their actual energy use is (some of this information is available through a load study), and not knowing who is going to participate in the program and what they actually will do under the program. (This information is only available through load evaluation.)

A utility may want to increase electricity sales by promoting heat pumps in gas heat areas. Money is spent on advertising and rebates are offered as incentives. The same advertising dollar spent on certain customers will cause them to make the investment, while others will not respond at all. Of those that purchase heat pumps and receive the rebate, some will have smaller homes and others will have highly insulated homes, both of which will use less electricity and cause lower sales for the same rebate dollars.

Different types of customers will agree to participate in a program under different types of influences. These customers will have different stocks of appliances and will have different patterns of use for their appliances. Also, certain programs will affect certain appliances for one customer, but not for another.

Clearly, unless something is understood about customer response to a program, it is nearly impossible to determine how many or who will decide to participate in a program. But more importantly, unless something is understood about customer appliance usage, it will be difficult to get customers whose participation would have the most desirable load impact response into the program. The maximum impact for the program dollars will not be achieved. Without understanding the load impacts achieved from load modification, a program cannot be targeted to ensure the "biggest bang for the buck."

When a program is designed, it is usually understood that there is a desired population to be targeted; however, it is not always understood that the information needed to target accurately, as well as to check the results of the targeting, can really only come from the results of load evaluation. Load evaluation of existing programs can guide the design of new programs, and load evaluation of an ongoing program enables those in charge to keep the program on track or alter it to better fit a changing energy situation.

The load evaluation of conservation programs yields valuable information for new load management and marketing programs, as well as for possible redesign and redirection of the existing program. The load evaluation of a conservation program can indicate which loads are easily modified, which loads customers can and are willing to change, and the size of that change. These evaluations indicate who can be expected to respond and the level of response to various incentives such as rebates, price incentives, and bill reductions. They can indicate what loads actually are, and identify profitable and unprofitable loads which need modification. Load evaluations can also identify programs whose positive benefits go beyond simply achieving the desired load impact; i.e., they can identify when the utility is gaining positive public relations or other benefits from a certain type of program.

Another class of program worth mentioning here is the customer service program (e.g., the Balanced Payment Program and the Automatic Payment System). These programs are not usually thought of as affecting load and were designed to benefit the utility's cash receivables situation as well as to promote public relations. Load evaluation of these programs would be valuable to determine whether they affect customers' response to price signals and changes as well as participation in other programs. Since the customers' awareness of their energy use and cost is strongly affected by these programs, they may not be as responsive to utility influence applied through vehicles such as load modification programs.

The cost of these programs in operation, as well as any cost in load impacts, could then be compared to the value of the increased likelihood of collecting the accounts receivable and the positive public relations impact to determine the cost-effectiveness of these programs. The information gained would also be valuable to other programs, both as additional load data and as an indication of other influences on load impacts.

In summary, load evaluation is more valuable to the utility than simply indicating how a program did, whether it was cost-effective. The information gathered from any type of program is an invaluable aid in predicting the cost-effectiveness of future programs and the future of the ongoing program. A load evaluation program helps utilities target their load modification programs toward customers whose participation will ensure the desired load impact and away from customers whose participation will not, and thus achieve maximum benefits from their programs.

The evaluation functions developed as part of the build-up of conservation programs discussed earlier will be an important part of the marketing organization, but will require some modification. Many utilities have developed an ability to evaluate the impact of their conservation programs. They have load monitoring programs, for example, to measure customer loads. Methods of sampling customers which allow extrapolation of results to the entire system have been developed. Many utilities now have staff members trained to estimate the degree to which the program, or program feature, influenced customers to take the intended action. Models and computer programs exist to assist in estimating the load impacts.

Most of these capabilities are still useful for marketing. Instead of estimating impacts of conservation programs, the effects of marketing programs will be estimated. However, the existing efforts must be redirected in some ways.

First of all, estimating impacts of marketing programs is generally more difficult than it is for conservation programs. Often, conservation actions involve a technology or measure (such as replacement of lighting, or insulation) for which an engineering estimate of the impact can be prepared. It is usually the energy impact which is of interest in such instances. For marketing programs, it is usually the participation decision which is critical. For example, what is the relationship between TV advertising and the number of customers purchasing heat pumps? How many customers would have made the purchase even without the promotional program?

While this adoption decision has been involved in some conservation program evaluations (for example, in determining the effect of offering a zero interest loan for residential home weatherization), it has generally not been the focus of the analysis. With marketing programs, the adoption or participation decision generally is the focus.

In addition, evaluation is generally more important for marketing than it is for conservation programs. Often, the driving element of the conservation program was not the cost-effectiveness of the action, but the regulatory mandate behind it. Marketing programs must be justified on a cost-effectiveness basis, or it is likely that they will not be supported by commissions. No one wants a repeat of the "gold medallion home" era.

Finally, a major thrust behind marketing efforts today, at least in many parts of the country, involves a particularly difficult-to-measure factor. Many utilities are developing programs specifically targeted at retaining larger customers. These customers are seen as having a significant number of attractive energy service choices. In order to prevent (or slow down) the increasing trend toward bypass, utilities are offering discounts to these customers, through negotiations or via discount options such as interruptible rates. The key question is the viability of the threat these customers make when they discuss leaving the system.

Thus, while conservation-oriented evaluation efforts can strongly support marketing programs, they generally require an increased focus upon understanding the customer decision-making process. This requires not only more sophisticated modeling techniques, but also better integration with load and market research groups, to ensure that the necessary data are available.

RATE STRUCTURE IMPLICATIONS

While it is clear to most planners that a utility's rate strategies and its demand-side management (DSM) program choices are interrelated, few realize the full extent of this relationship.

Therefore, the transition from a conservation-oriented company to one with a marketing focus generally ignores the importance of the set of rate structures offered. This section is intended to demonstrate the importance of having the "correct" set of rates for the success of demand-side management programs.

To accomplish this demonstration, a set of DSM programs representing different load shape impacts will be examined for a hypothetical utility. The effect upon the cost-effectiveness of each program will be analyzed with four different rate structures: an inverted block rate, a declining block rate, a time-of-use (TOU) rate, and rates equal to marginal cost. Although many other rate forms are possible, this analysis yields some startling conclusions about the relationship between rates and demand-side management programs.

First consider the marginal costs of this illustrative utility. These are shown in Table 10-1. All of the marginal capacity costs are allocated to the summer or winter peak seasons, indicating the necessity of adding capacity should load grow in these periods. There is about a 2:1 variation in energy costs between periods as well.

Table 10-1. MARGINAL COSTS

Period	Capacity ($/KW)	Energy (c/kWh)
Summer Peak	96.00	8.0
Summer Off-Peak	--	4.0
Winter Peak	24.00	7.0
Winter Off-Peak	--	4.0

Each of the following types of residential programs will be analyzed:

- A conservation program, such as insulation

- A marketing program, such as promotion of heat pumps
- A peak clipping program, such as air conditioning cycling
- An energy shifting program, such as thermal storage
- A valley filling program, such as off-peak water heating.

A cost-effectiveness analysis is performed for each of these five load shape impacts in terms of each of the four rate structure cases outlined. It should be noted that each of the rates was designed to be revenue neutral -- the revenues produced by each are equal. It is instructive to note the relationship between each of the rates and the utility's marginal cost. The inverted rate has a tail block above marginal costs, with the first block below marginal cost. The declining block case is just the opposite; the tail block is below marginal cost, while the first block is therefore above it. Finally, the TOU rate is based upon the average marginal costs in each of the four periods; it does not go as high as marginal cost during the summer peak period, nor as low as marginal cost during the winter off-peak.

To evaluate each DSM program, a cost-effectiveness analysis is performed. To simplify the example, the costs of implementing each program are ignored. The impact upon rate level perspective is calculated. With this measure, the benefit of programs which reduce usage (i.e., conservation and load management programs) is the decrease in costs resulting from the fact that the loads which must be generated are reduced; i.e., the benefits are the change in load times the system's marginal cost. The cost of the program is the revenue loss, or the change in loads times the rate.

These benefits and costs are reversed for marketing programs. In these, the benefits are the increased revenues resulting from the new sales produced by the program, or the change in sales times the rate. The costs are the increase in costs brought about by the increase in sales, or the change in loads times the marginal cost.

Thus, the rate enters the calculation as part of the revenue loss for load reducing programs, and as part of the benefit for marketing programs. With this in mind, consider the costs and benefits of each load shape change.

As Table 10-2 shows, the benefits of conservation are not affected by the rate structure; the savings are simply the change in loads times the marginal cost. However, the costs of such a program are strongly influenced by the rate. With an inverted rate -- the tail block above marginal costs -- all sales reductions reduce revenues faster than they reduce costs. Therefore, the program will have the effect of raising the average level of rates. With a declining block rate -- the tail block below marginal costs -- the reductions in revenue will be less than the reduction in costs, thus making the program attractive. With rates equal to marginal costs, there will be a zero net benefit; the revenue loss must always just offset the savings from the program. Finally, with TOU rates which represent "average" marginal costs, the conservation program (which reduces sales uniformly across the year) will also reduce revenues by the same amount of the reduction in costs.

Table 10-2. IS CONSERVATION ALWAYS DESIRABLE?

Rate	Benefits	Costs	Net
Inverted	$6995	$9337	($2342)
Declining	6995	4669	2326
Marginal Cost	6995	6995	0
TOU	6995	6995	0

Consider the next situation for marketing programs, illustrated in Table 10-3. As can be seen, now the situation is reversed. The costs of the program do not change as the rate structure is changed, but the benefits do vary. With an inverted rate, the benefits must exceed the costs, since the tail block is above marginal costs -- thus, increases in revenues will come faster than increases in costs.

However, if the rate is declining block, the increase in revenues must be less than the increase in cost. Again, since the impact does not vary across the periods, the net benefit must be zero with marginal cost rates, and with the TOU (average marginal cost) rate.

Table 10-3. IS MARKETING DESIRABLE?

Rate	Benefits	Costs	Net
Inverted	$9337	$6995	$2342
Declining	4669	6995	(2326)
Marginal Cost	6995	6995	0
TOU	6995	6995	0

Now consider what happens in a program with a temporal aspect. As an example, peak clipping is illustrated in Table 10-4. Again, the benefits of the program do not vary with the rate structure, but the costs do. The revenue loss must be greater with the inverted rate than with a declining block rate. Since this program reduces load during the period in which the marginal costs are the greatest (the summer peak period), the benefits will be greater than the revenue loss for every rate except the full marginal cost rate.

Table 10-4. PEAK CLIPPING

Rate	Benefits	Costs	Net
Inverted	$2110	$ 831	$1279
Declining	2110	416	1694
Marginal Cost	2110	2110	0
TOU	2110	1247	863

What about an energy shifting program? In this case, as illustrated in Table 10-5, the benefits again do not vary with the rate structure. Since the program works by shifting energy from one time to another without changing the overall level of sales, there would be no revenue loss for either the inverted or declining block rate. Again, the revenue loss must equal the savings with the marginal cost rate, and be not quite as great for the TOU rate. In the latter case, this is because the rate is not quite as great as marginal cost during the peak period (when energy is being reduced), and is somewhat more than marginal cost during the off-peak period (when energy is being increased). Thus, the overall effect is to reduce costs somewhat more than revenues are reduced.

Table 10-5. ENERGY SHIFTING

Rate	Benefits	Costs	Net
Inverted	$2326	$ 0	$2326
Declining	2326	0	2326
Marginal Cost	2326	2326	0
TOU	2326	1138	1188

Finally, the results of a valley filling type of program are shown in Table 10-6. Since this is a program which increases load, the costs (the increase in costs associated with the new load) will not vary with changes in the rate structure. However, the benefits do change. With inverted rates, the increases in sales each bring in revenue which is greater than the marginal cost. These increases are substantially less with declining block rates. Note that these increases in sales (and costs) occur during the period with the lowest marginal costs. If this were a "peak building" program, the net benefit would be negative with the declining block rate. With rates equal to marginal costs, the gains just offset the losses. Finally, the benefits are just greater than costs with the TOU rate.

Table 10-6. VALLEY FILLING

Rate	Benefits	Costs	Net
Inverted	$7674	$3837	$3837
Declining	3837	3837	0
Marginal Cost	3837	3837	0
TOU	4820	3837	938

These results can be summarized by making the following four points. First, rates which are set equal to marginal cost can capture many of the benefits of any load shape program. Such rates ensure that changes in loads are balanced by changes in costs. However, in practice, it is very unlikely (one would like to say impossible) for rates to just equal marginal cost. It is quite difficult to know a utility's marginal costs exactly. Moreover, these costs will change over time, so that even if marginal costs and rates are initially set exactly equal, they will soon differ. Also, the accounting cost revenue requirement means that marginal cost rates will either collect too much or too little revenue to allow achievement of the target rate of return. Finally, during some times of the year (e.g., the summer peak hours), it will be quite difficult for many utilities to set rates as high as full marginal cost.

Second, the attractiveness of DSM and marketing programs varies strongly with the rate structure. The clear implication is that conservation programs should be put into place in situations where rates are below marginal cost, and marketing programs emplaced where rates exceed marginal cost. This strategy is only common sense; it requires that one simply build load in profitable markets, and reduce sales where losses are accruing.

Third, DSM and marketing programs which appear undesirable can be made attractive by changing the rate structure. While this prescription can obviously lead to foolish actions, it is clear that the cost-effectiveness of any load shape change cannot be evaluated in the absence of the rate strategy. Changes in one must lead to changes in the other. Since it would not make sense to change rate structures just to improve the cost-effectiveness of DSM programs, it follows that both rates and DSM programs must be designed together.

Finally, the trade-offs inherent in the evaluation must be recognized. Developing rates which track system costs better will lead to an improved selection of DSM programs, and thus long-run efficiency for the utility, but it will also lead to other impacts as well. Some customers will be helped by such a change, while others will be hurt. The stability of revenues to the utility could be affected. The amount of class revenue subsidization could be changed. These impacts should also be considered.

The primary guiding principle must be to develop both rates and load shape strategies together. The greater the extent that marginal costs can be utilized in the rate design process, the better the selection of DSM programs that address needs of the utility will be. Finally, when utility planners recognize the need for a change, a transformation from a conservation to a marketing orientation, it is a clear signal that rates must also change. Otherwise, not only will "wrong" price signals be sent to customers, but the wrong kinds of DSM programs will be selected and offered to customers.

SUMMARY

Changing market conditions are prompting utilities across the country to initiate aggressive energy marketing programs. While marketing may initially seem to conflict with existing conservation efforts, each activity is an appropriate element of a long-term corporate load shape strategy. Moreover, conservation departments, systems, and tools within a utility not only offer a useful foundation for new marketing activities but also contribute significant noneconomic benefits to utilities in a marketing mode.

Organizationally, marketing activities should be consolidated and responsibility for marketing placed at a high level within the organization. Reeducation of both employees and ratepayers is critical to full implementation of new marketing efforts.

New marketing activities tend to be highly selective; thus there is an increased need to understand the customer decision-making process and resultant load impacts. Evaluation functions developed for conservation programs can be augmented with more sophisticated modeling techniques and better integrated with market research groups to ensure their effectiveness in marketing areas.

Finally, rate schedules are a significant tool for achieving load shape objectives and can be designed to promote both conservation and strategic marketing. Rates based on marginal costs will accurately signal prices to consumers as well as enable the utility to select and offer the most cost-effective mix of conservation and marketing programs.

REFERENCE

[1]Edison Electric Institute, Case Studies in Marketing Organization, Volume 8, 1984.

CHAPTER 11

Electric Utility Marketing Programs

Clark W. Gellings
Todd D. Davis
Dilip R. Limaye

INTRODUCTION

The electric utility industry is embarking upon a new era. Distinguishing this era is the recognition that electric utilities are becoming increasingly diverse in terms of their system characteristics, load shape objectives, markets served, competition, customer mix, and a host of other factors. In addition, it is becoming more common for electric utilities to consider demand-side solutions to what have traditionally been viewed as supply-side problems.[1]

For some utilities, the risks that accompany capacity expansion may outweigh the lower risks of demand-side programs. However, before utility managers can plunge headlong into designing and implementing marketing programs, they need comprehensive information on the following aspects of marketing program planning, analysis, and implementation:

- Customer acceptance of alternative program designs and promotional efforts
- Customer response to, or load shape changes resulting from, implementation of specific programs
- Methods for consolidating bundles of technologies and promotional activities to increase the effectiveness of those activities

- Planning approaches and processes useful for designing marketing programs

- Research inputs used to design programs

- Utility experiences with program implementation, both successes and failures

- Contents and structure of marketing plans

- Evaluation and measurement of marketing program impacts.

The changing utility planning environment has led many utilities to implement programs that influence customer load shapes in ways that will benefit both the customer and the utility. Utilities are exploring demand-side management (DSM) options for valley filling and selective load growth in addition to load management and strategic conservation. A common objective of many programs is load factor improvement, with resulting benefits to the utility program participants, and even nonparticipants.

THE EPRI PROJECT

A number of utilities have experience in the design, marketing, operation, and evaluation of demand-side management programs intended to improve load factor. In order to disseminate the valuable information gleaned from these programs throughout the industry, the Electric Power Research Institute (EPRI) undertook a study of 11 utilities. The study team worked with several utility managers active in marketing demand-side programs for load factor improvement. Eleven utilities with DSM programs were surveyed and asked to identify critical implementation issues and problems.[2] The utilities were also asked to share the processes by which they linked marketing programs to more general utility goals, and how they evaluated alternative DSM program designs and organized their internal resources.

This chapter summarizes the results of the study, describing the issues that surround the implementation of marketing programs, with emphasis on both strategic and tactical considerations. Although implementation methods are influenced by a number of utility-specific factors, the information in this chapter can be quite useful to utilities considering a variety of marketing programs.

A list of the case study utilities participating in this project is provided in Table 11-1.

Table 11-1. CASE STUDY UTILITIES PARTICIPATING IN EPRI PROJECT

Utility Name	Abbreviation Used in Chapter
Alabama Power Company	APCO
Baltimore Gas and Electric Company	BG&E
Commonwealth Edison Company	Com Ed
Iowa Power and Light Company	IP&L
Kansas Power and Light Company	KPL
Kentucky Utilities	KU
Lincoln Electric System	LES
Minnkota Power Cooperative	MPC
Ohio Edison Company	OE
Otter Tail Power Company	OTPC
Philadelphia Electric Company	PECO

Source: Synergic Resources Corporation[2]

FRAMEWORK FOR ASSESSING UTILITY MARKETING PROGRAMS

Utility marketing programs have used a wide range of methods designed to influence customer adoption, which can be broadly classified into six categories:[3]

- Customer education
- Direct customer contact
- Trade ally cooperation
- Advertising/promotion
- Direct financial incentives
- Rate incentives.

The selection of a particular market implementation method or a mix of methods depends on a number of factors, including:

- The utility's overall marketing philosophy
- Prior experience with similar programs
- Estimated program benefits and costs to the utility and the customer
- Stage of buyer readiness
- Barriers to customer acceptance.

Table 11-2 illustrates the major functions, advantages, disadvantages, and best applications of these six market implementation methods.

The primary purpose of each of these market implementation methods is to influence the marketplace and to change customer behavior. Utility managers must choose the method(s) that will obtain the desired customer acceptance and response. Customer acceptance refers to willingness to participate in utility DSM programs, adopt the desired fuel/appliance choice and efficiency level, and change consumption behavior. Customer response is the actual load shape change that results from customer action, combined with the characteristics of the devices and systems being used. Customer acceptance and response are determined by a number of decisions that affect the choice of the fuel and/or the appliance, the efficiency of the appliance or equipment selected, and the utilization patterns of the appliance/equipment. These decisions are influenced by demographic characteristics of the customer such as income, knowledge and awareness of the technologies and programs available, and decision criteria such as cash flow and perceived costs and benefits, as well as attitudes and motivations. The decisions are also influenced by other external factors such as economic conditions, energy prices, technology characteristics, regulation, and tax credits (Figure 11-1).

Selection of an appropriate market implementation method should be made in the context of an overall market planning framework[4] (Figure 11-2). Major elements to consider in selecting the appropriate marketing mix are:

- **Market Segmentation** -- Based on the load shape modification objectives, information on customer end uses and appliance saturation, and other customer characteristics (from consumer research). The market can be broken into smaller homogeneous units so that specific customer classes are targeted.

- **Technology Evaluation** -- Based on the applicability of available technologies for relevant end uses and load shape objectives. The alternative technologies are evaluated and the profitability of specific appliances assessed.

Table 11-2. FUNCTIONS, ADVANTAGES, DISADVANTAGES, AND BEST USES OF VARIOUS MARKETING MECHANISMS

Mechanism	Functions	Advantages	Disadvantages	Best Uses
Customer Education	Inform customers of program availability/eligibility	Reaches large number of customers at low cost per customer	Generally has relatively low impact	Increase customer awareness and interest in demand-side options
	Present advantages of technology or behavior being promoted	Provides broad coverage, keeps utility name and message before the customer		Present information on technology, program benefits, and availability/eligibility requirements
Direct Customer Contact	Allow for more exact and customized analysis of alternatives, and installation of demand-side options	Allows customized service and in-depth marketing; good for closing the sale	Tends to be quite costly per customer contact	Consultation with customers regarding large expenditure purchases helps reduce risk and uncertainty
		Personalizes the utility/customer relationship	Can engender negative reaction from trade allies	
Trade Ally Cooperation	Encourage purchase of electric technologies/appliances at point-of-sale and increase the channels of distribution	Utility does not bear all costs of marketing effort	Requires coordination	General promotion of appliances where point-of-purchase advertising/persuasion is likely to be beneficial
		Utility gains interested partner in the marketing effort	Some loss of program control	
		Third party endorsement		
		Greater coverage during critical decision periods		
		Provides after-sale service resource		

Advertising/ Promotion	Increase program awareness Present advantages of electric technologies or behaviors being promoted Move customer toward adoption	Broad or specialized targeting possible, depending upon media chosen Keeps utility name or program concept before customers Can present utility in similar format as other purveyors of service to the customer	Can be expensive Impact generally difficult to measure	Program announcements, specialized presentations of technologies/ benefits
Direct Financial Incentives	Move customer to adoption of electric technology being promoted by appealing to their implicit discount rate	Can provide powerful incentive Reduces first costs to increase competitive appeal of electric technology Simple to administer	Can be expensive Generally needs to be "sold" to regulatory agency Can create negative reactions among customers and trade allies	Appliances with widespread applicability and easily definable benefits to the system as a whole
Rate Incentives	Move customer to adoption of electric technology being promoted by appealing to their implicit discount rate	Can provide powerful incentive to customer Reduces operating costs to increase competitive appeal of electric technology Utility capital shared with customer as benefits are received	Generally needs to be "sold" to regulatory agency Can create negative reactions among ineligible customers	Encouragement of technologies which clearly represent a definably distinct pattern of use (i.e., rate class)

Source: Synergic Resources Corporation[2]

Figure 11-1. FACTORS INFLUENCING CUSTOMER ACCEPTANCE AND RESPONSE

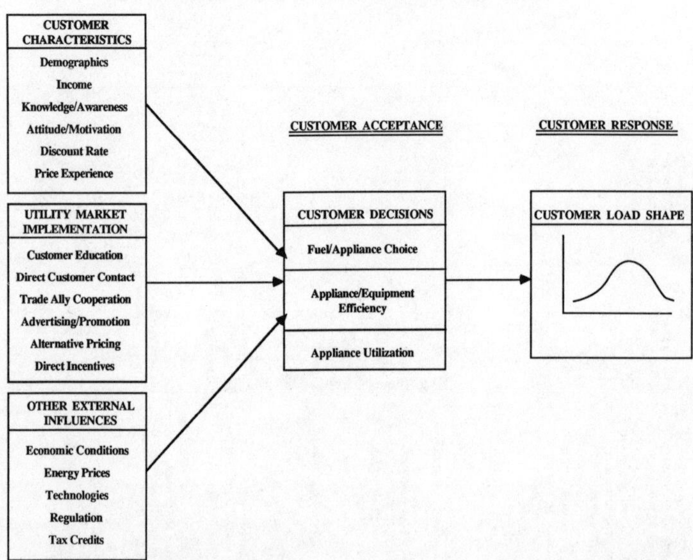

- **Market Share Analysis** -- Based on estimates of customer acceptance, this procedure calculates the proportion of the total potential market that can be served competitively by a utility.

- **Selection of Market Implementation Methods** -- Based on the above analyses and estimates of potential customer acceptance and response, the appropriate mix of implementation methods is evaluated and selected.

- **Market Implementation Plan** -- Based on the selection of specific market implementation methods, an implementation plan is developed to define and execute the DSM programs.

- **Monitoring and Evaluation** -- The results of implementation are monitored and evaluated to provide relevant information to improve future programs.

Figure 11-2. ELECTRIC UTILITY MARKET PLANNING FRAMEWORK

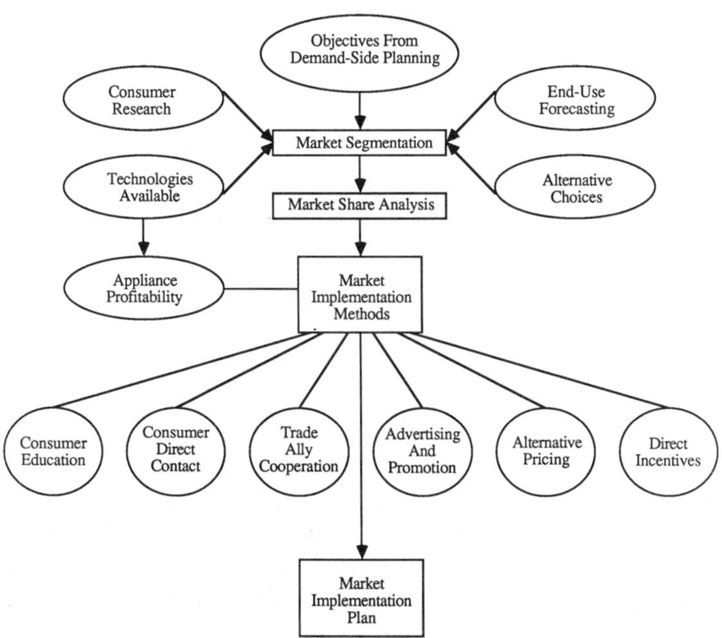

Source: C. W. Gellings and D. R. Limaye[4]

Methods used for market segmentation and target marketing can vary depending on the customer characteristics and technologies/end uses being addressed by the DSM alternative.[5] If a certain technology offers significant benefits to the customer and there is little or no perceived risk to its use, the technology is likely to be well accepted and there will be little need for the utility to intervene in the marketplace. However, if customer acceptance is constrained by one or more barriers, the market implementation methods should be designed to overcome these barriers. Barriers to customer acceptance may include:

- Perceived or actual return on investment (ROI)

- High first cost
- Lack of knowledge/awareness
- Lack of interest/motivation
- Decrease in comfort/convenience
- Limited product availability
- Perceived risk.

Figure 11-3 shows how the market implementation methods can apply to overcoming the barriers.

A second important consideration involves buyer readiness. Customers generally move through various stages toward a purchase decision, and each customer stage has a bearing on the appropriateness of the market implementation methods used (Figure 11-4). In most instances, effective research can identify where customers are in their decision process. Key questions to consider are:

- Do customers perceive a need to control the cost of energy, and are they aware of alternative DSM technologies?
- Where do customers go to search for more information and guidance on alternatives, and what attributes and benefits are perceived for any given option?
- How much interest is there in participating in a DSM program, and how can customers be influenced to move toward participation?
- What specific attributes and benefits must customers perceive in order to accept a particular DSM technology?

Figure 11-3. APPLICABILITY OF MARKET IMPLEMENTATION METHODS TO OVERCOME BARRIERS TO CUSTOMER ACCEPTANCE

BARRIERS TO CUSTOMER ACCEPTANCE	MARKET IMPLEMENTATION METHODS					
	Customer Education	Direct Customer Contact	Trade Ally Cooperation	Advertising/ Promotion	Alternative Pricing	Direct Incentives
Low Return On Investment (ROI)					●	●
High First Cost, Favorable ROI	◒	◒		◒	◒	●
Lack Of Knowledge/ Awareness	●	●	◒	●		
Lack Of Interest/ Motivation		◒	◒	●	◒	◒
Decrease In Comfort/ Convenience	◒	◒			◒	●
Limited Product Availability		◒	●			
Perceived Risk	◒	●	●		◒	●

● Generally Applicable ◒ Conditionally Applicable [Blank] = Rarely Or Not Applicable

Source: Battelle-Columbus Division and Synergic Resources Corporation[3]

- How satisfied are customers who participated in previous DSM programs?

It must be reemphasized that market segments can be defined using a number of criteria, some of which may be interrelated. Segmentation of markets by end use, stage of buyer readiness, perceived barriers to acceptance, and other socio-demographic factors can suggest the appropriateness of alternative market implementation methods.

Figure 11-4. APPLICABILITY OF MARKET IMPLEMENTATION METHODS TO STAGE OF BUYER READINESS

STAGE OF BUYER READINESS	MARKET IMPLEMENTATION METHODS					
	Customer Education	Direct Customer Contact	Trade Ally Cooperation	Advertising/ Promotion	Alternative Pricing	Direct Incentives
Need Recognition/ Awareness	●	◒	◒	◒		
Search For Alternatives/Interest	◒	●	●	●		
Purchase/ Adoption		◒	●	◒	●	●
Satisfaction		◒	◒			

● Generally Applicable ◒ Conditionally Applicable [Blank] = Rarely Or Not Applicable

Source: Battelle-Columbus Division and Synergic Resources Corporation[3]

MARKETING PROGRAM EXPERIENCES FROM CASE STUDIES

Brief descriptions of the experiences of the case study utilities with each of the six market implementation methods are provided below. Most of the marketing programs reviewed in the EPRI study had been in the field for a relatively short time (less than two years), and therefore it was not possible to report the full impacts of these programs. Nevertheless, some important lessons can be learned from the experiences of the case study utilities.

Customer Education

As a utility marketing implementation strategy, the purpose of customer education is to convince customers of the benefits they can obtain through participation in utility programs or through certain energy technology or utilization choices (Figure 11-5). Although customer education can begin with simple awareness, its real point is persuasion. The customer education strategy generally places emphasis on explanation and exposition, and is often accomplished through direct mail techniques. In some cases, one mailing may include both advertising and customer education materials, as is the case in PECO's heat pump mailings. The package not only announces the program, but goes on to provide detailed information on heat pump performance, savings, and applicability. It even includes a means for taking the next step toward the heat pump purchase: a mail-back card to arrange a visit and bid from a heat pump installation contractor.

Figure 11-5. CUSTOMER EDUCATION PROGRAMS

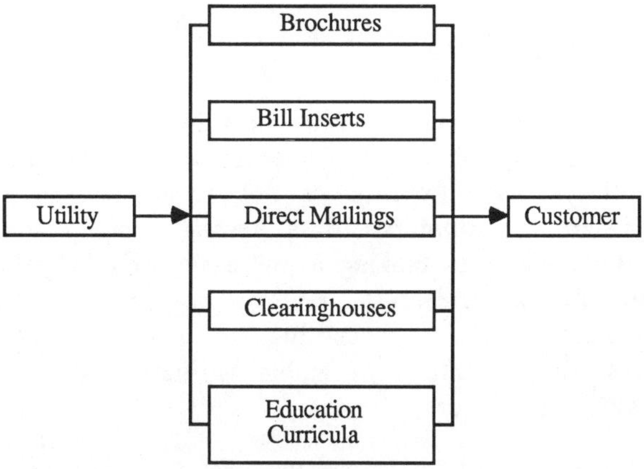

Source: Battelle-Columbus Division and Synergic Resources Corporation[3]

By a wide margin, case study utilities favored brochures and booklets as their main means of providing customer education services. Utilities prepare printed matter on a wide variety of topics ranging from low-cost, no-cost energy conservation hints to installation instructions for insulation and heat pumps. Several of the case study utilities provide targeted educational publications such as PECO's "Conserving Energy in Religious Institutions." Most of the case study utilities were found to offer a dozen or more customer education publications. These publications are often announced through bill inserts and made available upon customer request. They are also distributed at utility exhibits at local energy fairs, at homebuilder and other trade shows, and at special utility-sponsored topical energy seminars and workshops.

Case study utilities use a variety of customer education formats to reach various customer classes. Brochures and general publications were most often targeted to residential customers. Exhibits were used for both general public and trade audiences. Seminars and workshops were more often and most usefully offered to commercial and industrial customers, because they could more readily make use of detailed topical information.

Direct Customer Contact

Direct customer contact can provide a very personalized, flexible, and effective market implementation strategy (Figure 11-6). Because of the relative expense of this strategy, it is most often reserved for larger customers, those having more complex energy systems, or those closer to making a purchase decision. However, the exception to this general rule is the provision of energy audit services to homeowners. According to several utility company representatives, the advantage of audits is that they put the utility face to face with its customers.

Some audit activity, with similar direct impact benefits, was reported by case study utilities for commercial customers. Most of the utilities providing these services are doing so because they recognize that the commercial customer class is diverse and very important to their operation and profitability. This justifies the greater customer contact that is provided by energy audit services.

Figure 11-6. DIRECT CUSTOMER CONTACT OPTIONS

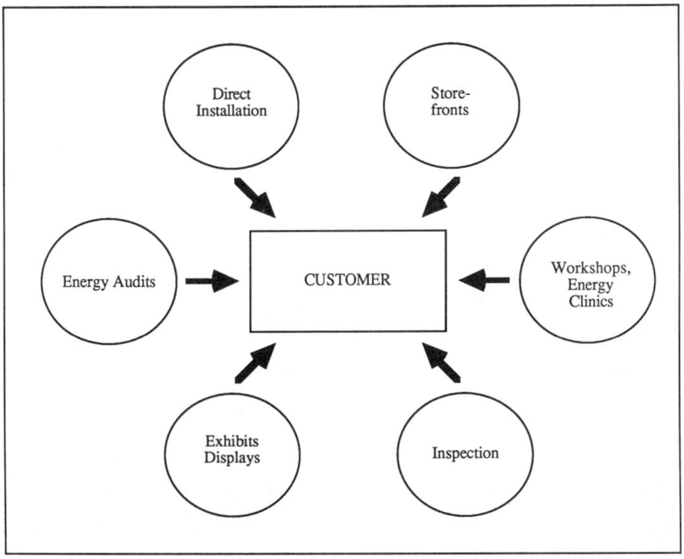

Direct Installation
Store-fronts
Energy Audits
CUSTOMER
Workshops, Energy Clinics
Exhibits Displays
Inspection

Source: Battelle-Columbus and Synergic Resources Corporation[3]

In at least half the case study utilities, individual representatives are given personal responsibility for specific large accounts, both industrial and commercial. It is in this form that direct customer contact and the personalization of the utility/customer relationship reaches its fullest realization. While usually careful not to infringe on the role played by energy management consultants or consulting engineers, the utility representative often becomes more than a one-time visitor. Instead, he becomes an available energy expert as well as the customer's one-stop point of contact for all matters related to utility service, from questions about bills to new service, from energy management to electrification opportunities.

Trade Ally Cooperation

A trade ally is defined as any firm, individual, or organization whose interactions with its customers or members can influence the relationship between the utility and its customers (Figure 11-7). Almost every case study utility has implemented a program in which the involvement of a trade ally has played a prominent role. However, these relations have not been altogether smooth.

Several of the case study utilities reported that when they first contacted their traditional trade ally groups as part of the renewed emphasis on marketing programs, the trade allies often expressed resentment at how abruptly and completely the utilities had abandoned them during the "conservation era." Fortunately, however, according to the same utilities, the trade allies were equally glad to see the reemergence of utility marketing efforts and have since helped ensure the success of several important demand-side programs.

Trade ally involvement in utility programs has not simply picked up where it left off in the late '60s and early '70s. More and more input from the trade allies is evident in the current trade ally programs. Several utilities reported that they had held discussion groups or focus groups with trade allies to fully explore how the tradespeople and the utility could best cooperate. One utility actually undertook a survey of a representative sample of potential trade allies in its service territory. Several utilities reported that they used to plan programs and then "announce" them to potential trade ally participants; now they develop only the program concept, then invite the trade allies in to discuss the details of program implementation and administration. According to these utilities, such practices have done more than simply assuage the feelings of trade allies -- they have helped ensure the success of the programs. For instance, when PECO invited heat pump installation contractors in to discuss its heat pump conversion and add-on heat pump programs, the trade allies suggested a revised timetable for the program that would enable them to deliver program services more promptly to customers.

Figure 11-7
ILLUSTRATION OF TRADE ALLY ROLE IN INFLUENCING CUSTOMER

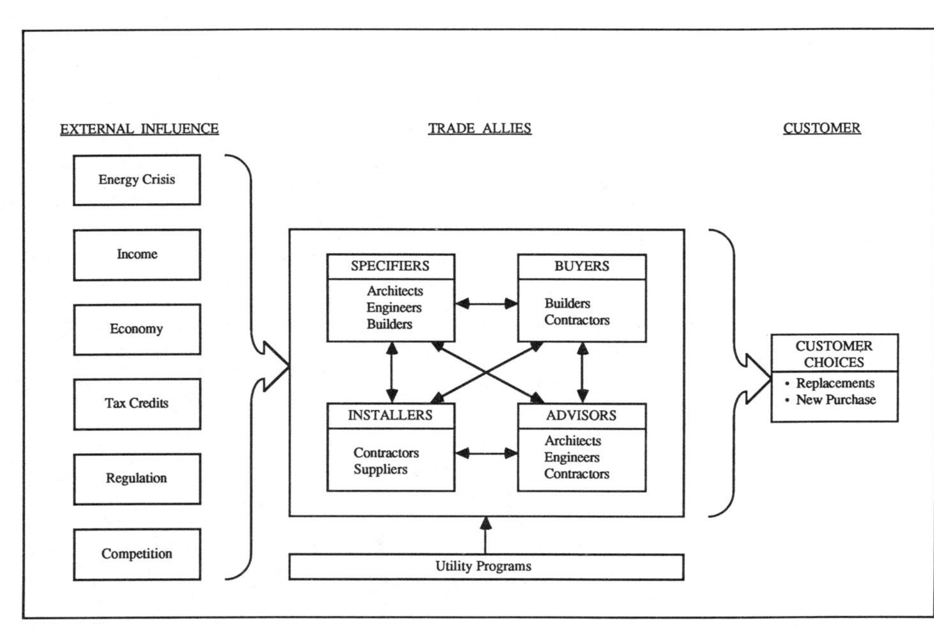

Source: Battelle-Columbus Division and Synergic Resources Corporation[3]

A key basis of trade ally involvement in any utility demand-side program is the potential for increased business for the trade ally. Most case study utilities offer further inducements as well. The most popular incentive provided by utilities is cooperative advertising, with newspaper ads being used most often. In addition to helping to pay for ad placement, some utilities provide technical advertising assistance to trade allies. MPC, for instance, has undertaken the design, layout, and printing of materials promoting the dual fuel heating program for its distribution co-ops. Other utilities have provided consulting services to trade allies in the design of co-op ads.

Other inducements for trade ally participation in utility demand-side programs include sales incentive payments and spiffs. BG&E pays heat pump installers and builders $10 for each card they send in to the utility listing the date, address, make, and size of a heat pump installation. In addition to providing a way to check that heat pump sales have actually taken place, this mechanism provides the utility with specific information valuable for market penetration analysis, customer follow-up, and load research. APCO, by contrast, relies on financial incentives to gain builder support in becoming part of the "Good Cents Home Program." Company representatives can give builders a wide variety of gifts including ball-point pens, business folders, baseball caps, travel mugs, and sports duffel bags.

Advertising/Promotion

The use of advertising and/or promotion to elicit customer response to utility demand-side programs follows the same rationale as it does in other industries. The customer must be aware of a product or service before he can choose to use it. The function of advertising is, first, to make customers aware of the existence and benefits of utility demand-side programs, and second, to influence customer preference toward these technologies and energy uses. In its most common form, advertising can be accomplished through bill inserts which announce the availability of a utility program. Another direct mail approach, but one that is much less often used, is targeted mailing. In this approach, customers are screened using certain factors (e.g., demographics, appliance saturation, energy use level) to identify those more likely to participate in the program. Although the screening requires time and resources, it results in a smaller, more homogeneous and targeted mailing list. Besides requiring less material and mailing costs, this list, due to its increased homogeneity, allows development of more pointed advertising themes.

As an example, PECO uses direct mail advertising in both the general and targeted manner. It sends bill stuffers announcing the availability of low cost energy audits to all its residential customers. By contrast, the company sends information packets on its heat pump program to only about 77,000 of its residential customers. In this case, the utility uses a targeted approach in which prime candidate customers are identified through an analysis of company billing records.

After direct mail, newspaper advertising is most popular with the case study utilities. Newspaper ads are viewed as the least costly form of general media advertising. Cooperative advertising with trade allies often relies on newspapers as well. Radio and TV ads are used less frequently. To a large extent, this is a result of the much higher price of radio and TV advertising, particularly in larger metropolitan areas. However, the use of these advertising media is increasing and reveals the growing sophistication of utilities in advertising. IP&L, for instance, has used a series of very well produced radio and TV ads which promote the use of heat pumps and customer awareness of the value of electric service. IP&L also purchases media time and selects newspapers and broadcast stations based on reach and frequency criteria.

These types of messages have been augmented by many of the case study utilities with corporate ads. In corporate ads, no particular program or service is promoted. Rather, the ad seeks simply to create a positive image of the company for the customer. Examples of utility corporate advertising slogans include PECO's "We see how much needs to be done," and Com Ed's "Doing things right -- and proud of it." Use of such ads by the utilities mirrors their increased use by many major U.S. corporations. In all cases, the purpose is to create a favorable impression of the company which, indirectly, should increase customer acceptance of the company's goods and services.

Direct Financial Incentives

Direct financial incentives can present the strongest inducement for customers to participate in utility marketing programs because they are designed to reduce the customer's net cash outlay or payback period. Examples of direct financial incentives include cash rebates, low interest loans, billing credits, and the provision of equipment and/or services, such as installation and maintenance. While these strategies can be quite successful in spurring market penetration, they can also be quite expensive for the utility using them. Because of their potential high cost, direct financial incentives have been used sparingly by the case study utilities. In some cases, utility representatives expressed negative attitudes regarding "giveaway" programs. Such programs, it was felt, do not require any commitment to program objectives or benefits on the part of participants and, therefore, often result in higher costs and lower impacts for the sponsoring utility.

While the most expensive direct financial incentives were generally avoided, case study utilities did use certain of these implementation strategies in a controlled manner in order to stimulate market adoption of efficient electric end-use technologies. Perhaps the most aggressive use of direct financial incentives was exhibited by MPC and its 1983 associated distribution cooperatives. At the outset of its 1983 dual fuel heating program, MPC told the distribution co-ops that it would give them $150 for each dual fuel system installed by their customers, and that the distribution co-op was encouraged to add $50 to the incentive rebate either as a cash rebate or as a means of providing a low interest loan. MPC had earlier provided a limited number of free dual conversion kits to the distribution co-ops as well. Later in the program, the incentive from MPC was converted to a rate incentive only (see discussion in following section), yet most of the distribution co-ops chose to continue their direct incentives to their customers. This, then, represented a rate incentive from the generation and transmission co-op to the distribution co-op, winding up in a direct financial incentive for ultimate customers.

OE, APCO, and LES also offer direct financial incentives. OE extends wiring allowances to employees who upgrade their homes for increased electric end-use service. While the amounts of the wiring allowances are not small, ranging from $300 for an add-on heat pump to $800 for total electric conditioning service, the number of persons eligible for the program is relatively small, thereby limiting total potential program costs. APCO offers an Assured Service Program which provides a maintenance service contract for homeowners who add heat pumps to their residences. The service is provided by private heat pump contractors but is underwritten by the utility. This program neatly solves two problems with one stroke; it directly addresses customer concerns about the service reliability of the new technology and does so in a way that lowers both the operating cost and perceived risk of using such a system. LES provides billing credits for those customers using off-peak domestic water heaters. The billing credits are combined with an off-peak rate to provide a very favorable competitive position for electricity in terms of serving water heating loads.

Rate Incentives

Indirect financial incentives can be supplied to customers through the utility's rate structure. The advantage of rate incentives to utilities offering demand-side programs is that they are relatively inexpensive and occur in phase with the benefits the utility receives from program implementation. In addition, to the degree that incentive rates are based on the marginal cost to serve specific energy end uses, it can be argued that they represent accurate economic signals to the customer regarding the value of shifting consumption to certain periods or purposes. In this sense, incentive rates can help provide a more realistic market price signal for the electric utility industry. This is the philosophy that forms BG&E's reliance on rates as the mainstay of their marketing strategy.

Company officials feel that, whenever possible, the customer should receive a signal from the rate structure as to the most efficient energy utilization choice. The company routinely investigates each demand-side technology for the applicability of a special rate.

Over half of the case study utilities reported some use of rates as a pricing mechanism rather than strictly as a revenue requirement mechanism. This, in and of itself, reveals the renewed importance of marketing in the case study utilities. Furthermore, several of the case study utilities reported that, as a result of recent reorganizations, their rate departments are now part of their marketing departments, thereby further integrating the pricing function as part of their demand-side planning activities.

Examples of innovative rates include time-of-use rates, seasonal rates, inverted rates, curtailable or interruptible rates, and certain special off-peak or load controlled rates. Most of the case study utilities offered at least one of these types of rates. Time-of-use rates were the type of incentive rate most popular with the case study utilities and tended to be available to all or most customer classes. Curtailable and interruptible rates were also offered with considerable frequency, especially to larger industrial and commercial customers. The MPC dual heat program represents one of the few load controlled interruptible rates available to residential customers.

A new type of incentive rate that has been adopted by several of the case study utilities is the economic development rate. The basic reasoning behind economic development rates is that the profitability of the utility is tied to general economic conditions in two important ways. First, better economic conditions mean more people are working and earning more money. Therefore, they will have the ability to pay utility bills and to purchase and use discretionary energy-using appliances. Second, better economic conditions mean increased industrial and commercial loads. As a result, several of the case study utilities, including OE and PECO, have introduced economic development rates.

SUMMARY OF MARKETING PROGRAM RESULTS

In most cases, utility marketing programs have been quite successful in reaching their goals. Generalizations that can be drawn from the implementation experiences of the case study utilities include:

- Heat pump programs have had their greatest success in new construction (either single- or multifamily), in rural areas, and in suburban areas where natural gas is unavailable.

- The market for whole house heat pump conversions and add-on heat pumps is much more difficult to penetrate than the new home market. Reasons include the first cost bias of homeowners and the need to market individually with homeowners for each unit placed, rather than marketing to builders for multiple installations. Customers also do not understand that add-on heat pumps can be installed on existing heating systems.

- Few utilities have marketed heat pump water heaters in the past. Some utilities are now starting to promote them.

- Employee incentive programs appear to be effective in stimulating marketing activity and increasing customer acceptance. While no formal evaluations of these programs have been done, utility personnel feel that they have made a difference; one utility attributes 10% of its incremental sales to the effects of the incentive program.

- Trade ally involvement is seen as extremely important to the success of marketing programs.

- Multiple marketing methods were often necessary to achieve program penetration goals.

- Marketing to industrial and large commercial customers based on intensive personal contact and incentive rate structures (e.g., interruptible rates) has been very successful.

- Significant opportunities exist to design effective marketing programs in all customer classes, but especially in the commercial customer sector.

MPC, one of the few case study utilities that has had a demand-side program in place for over five years, has been able to gauge the effect of its efforts. In part this is due to the longevity of the program; in part it is due to the fact that the program promotes load controlled residential space heating, the total system effect of which can be easily ascertained by throwing a switch. MPC officials have concluded that the program has:

- Improved annual load factor 15 percentage points, from 48% in 1976 to 63% in 1983

- Maintained electricity rates at a level at least 36% below what they would have been in the absence of the program.

Though using shorter time frames, other utilities claimed similarly favorable results from their marketing programs:

- APCO reached or exceeded virtually all of its 14 program goals for 1983. Several of the goals were exceeded by considerable margins; for instance, peak demand reductions of 1,932 kW were achieved, 238% of the 812 kW goal.

- BG&E far exceeded its goals for incremental off-peak space heating loads in the industrial and commercial sectors. Saturation of heat pumps in new residential construction has reached such high levels due to successful competitive pricing strategies that formal program efforts have been discontinued.

- Com Ed continues to get all of the new load represented by new building construction in Chicago's Loop. The utility credits aggressive field representatives and appropriate end-use systems for this success.

- IP&L credits its marketing efforts with having achieved 55,000 MWH in sales and $2.5 million in revenues over and above company projections for 1983.

- KPL reports 1,923 additional heat pump installations in less than two years as a result of its marketing efforts.

- LES more than doubled the number of new electric heat connections in the first year of its marketing program; response to voluntary programs involving air conditioning load management, commercial off-peak rates, and industrial interruptible rates has exceeded company expectations.

- OE exceeded all goals for incremental load and consumption in the commercial and industrial customer classes; incremental sales of 599,373 MWH and new load management of 65,343 kW reported.

- OTPC has enlisted between 300 and 400 heating and electric contractors as trade allies in its programs, placed over 350 customers on demand control in 1983 alone, and credits its 1983 marketing programs with adding 128,531 kWh in sales.

- PECO reported a 4% increase in total sales, attributable in large part to marketing efforts. Heat pump penetration in new construction was particularly high, reaching 64% of all new residences. Area development results were also described as exceeding expectations.

Of course, the case study utilities also experienced areas in which achievements fell short of expectations. As mentioned earlier, this was most often the case regarding residential retrofit technologies, such as add-on heat pumps. Perhaps more important was the emerging trend observed by several case study utilities in which competitors offer programs in direct response to and in direct competition with those of the electric utilities. While it is not surprising that such programs would be developed and offered, their emergence underscores the increasingly competitive environment in which electric utilities must operate. Furthermore, it requires that electric utility responses to competitive conditions remain flexible in order to maintain and improve the industry's market share during this period of market volatility. In short, while the early rounds of electric utility marketing efforts have been successful, utilities perceive the need to continue and expand their marketing efforts in innovative and creative ways in order to protect and extend their market position.

REFERENCES

[1]Battelle Columbus Division and Synergic Resources Corporation. Demand-Side Management: Overview of Key Issues, Electric Power Research Institute, EA/EM-3597. Palo Alto, CA. 1984.

[2]Synergic Resources Corporation. Marketing Demand-Side Programs to Improve Load Factor, Electric Power Research Institute, EA-4267. Palo Alto, CA. 1985.

[3]Battelle-Columbus Division and Synergic Resources Corporation. Demand-Side Management: Technology Alternatives and Market Implementation Methods, Electric Power Research Institute, EA/EM-3597. Palo Alto, CA. 1984.

[4]Gellings, Clark W. and Dilip R. Limaye. "Market Planning for Electric Utilities." Paper presented at the Energy Technology Conference, Washington, DC, March 1984.

[5]Lloyd, Gayle and Todd D. Davis. "Identifying Commercial/Industrial Market Segments for Utility Demand-Side Programs." Paper presented at the Great PGandE Expo, Oakland, CA, 1986.

CHAPTER 12

Marketing Strategies for Rural Electric Cooperatives

Clark W. Gellings
Todd D. Davis
Richard Tempchin

"Electrical consumption is the result of someone's marketing effort."

--Evangeline W. Jackson,
Chairperson of the Marketing Committee, Virginia,
Maryland and Delaware Association of Electric Cooperatives

INTRODUCTION

From their beginnings in 1937, rural electric cooperatives (RECs) have had to market their product. Farmers had to be taught how to use electricity; the cooperatives had to advertise and actively promote the sale of electric appliances. Cooperatives served as both dealer and service center for electric irons, refrigerators, stoves, washing machines, and water heaters. This original marketing campaign had one simple objective -- to sell more electricity, for economies of scale made one-cent per kilowatt-hour (kWh) electricity possible. Kilowatts could be generated so cheaply that some even questioned the need to meter consumption. Fifty years, eight administrations, three wars, two oil crises, and many environmental and regulatory acts of Congress later (not to mention Three Mile Island, Chernobyl, and scattered minor nuclear accidents), RECs still have to market their product, but in a vastly different manner.

Today's utility marketing objectives are varied, complex, and antithetical to those of the early days. Wildly fluctuating oil prices, the enormous costs of nuclear and other new capacity, and stringent environmental regulations have redirected marketing efforts. Instead of merely building load, current load shape objectives include peak clipping, valley filling, load shifting, strategic conservation, strategic load growth, and flexible load shape.

The industry and Congress now regard demand-side management as a bona fide resource, and marketing by means of demand-side management programs dominates utility market planning. This chapter will focus on the special needs and problems of RECs relative to demand-side marketing strategies, drawing upon the experiences of the Virginia, Maryland and Delaware Association of Electric Cooperatives and several other generation and transmission cooperatives to illustrate various points.

THE EFFECTS OF SOCIAL CHANGE

Today's electric cooperative membership is much different from that of earlier years. Members are exposed to substantial amounts of information from regional market areas. Families are much more mobile than ever before. Young people from rural families are moving to more urban areas for better education and jobs. Once isolated, small, slowly growing rural communities located within a short drive of larger urban areas are now encountering suburban sprawl. This results in a much more urbanized household living in what was once a primarily rural community.

The REC member of today is less likely to be as loyal and informed about rural electrification and its philosophy as were previous members. To the current member, the utility may be viewed as a commodity broker, rather than as a non-profit electric service organization.

These social changes suggest that the nation's rural electric cooperatives must strive for better knowledge and understanding of the values and needs of their members. The benefits that electricity provides today are far greater than ever before, yet loyalty and member satisfaction are declining. Many of the technological and life-style advancements that have taken place since the mid-1930s are now taken for granted and viewed as necessities. In the past, the aggregate and unrestricted use of electric technologies benefited cooperatives by lowering average costs. Now, average costs are increasing, establishing the need to find ways to manage load growth. The objective of the first 40 years of rural electrification was to improve the standard of living of rural households and farmsteads by simply getting the power to the point of use. A key issue for the next 15 years will be to determine how to add value to member electricity use by managing electric loads.

The need to manage member electricity use is tied directly to the desire to improve the efficient use of capital. Because of the capital-intensive nature of the electric utility industry, there is a significant need to control rates by limiting capital outlay for new generation facilities. Additional load growth can also lead to marginal cost increases in the areas of transmission and distribution. Should unrestrained price increases to members continue, further declines in member loyalty will result and the search for substitute energy sources will intensify.

DEMAND-SIDE MANAGEMENT AND RURAL ELECTRIC COOPERATIVES

The concept of demand-side management (DSM) centers on a partnership between the cooperative and its members. The cooperative designs the program -- in the best case, with substantial member input -- and offers it to its members, who ultimately decide whether to participate.

While DSM is not a cure-all for new industry challenges, it does provide utility management with many additional problem-solving alternatives that are applicable to a host of cooperative activities. For cooperatives with strong load growth, for example, load management and strategic conservation represent effective means to reduce or postpone construction or purchase of a new generation source. For cooperatives that purchase power under ratchet clauses or other rates that penalize purchases during periods of the seller's peak demand, DSM can reduce costs of purchased power. For other cooperatives, electrification and deliberate increases in the market share of energy-intensive uses can improve the cooperative load characteristics and optimize the reserve margin or power pooling arrangements.

Aside from these rather obvious cases, changing a cooperative's load shape can reduce operating costs. It can also permit adjustments in plant loadings, thus increasing the use of more efficient plants and permitting the use of locally abundant, less expensive energy sources.

Perhaps the greatest benefit of DSM, however, is the increased flexibility that it provides. As energy markets become increasingly competitive, structural changes significantly affect the U. S. economy, new end-use technologies and consumer-owned generation facilities penetrate the market, and other factors arise which make balancing demand and supply extremely difficult. Because of the inherent "lumpiness" of central station generation and the substantial cost that is often associated with construction schedule changes, supply-side options are generally unable to provide needed planning flexibility in the face of load forecast revisions. DSM activities, however, are generally much more amenable to relatively short-term changes. In fact, some DSM activities (e.g., direct load control) are specifically designed to provide nearly instantaneous responses that match demand to available generation, transmission, and distribution resources.

CONSUMER MEMBERS OF A RURAL ELECTRIC COOPERATIVE

DSM alternatives give consumer members of a rural electric cooperative a positive signal that their cooperative is concerned about high electric bills and is providing its consumers with the means to control future expenditures for electricity. Cooperatives are in a unique position to employ successful DSM programs thanks to their relationship with their consumer members. As the rural-electrification movement began, groups comprised largely of farmers organized to form cooperative organizations. The success of the movement was rooted in the establishment of a relationship between consumers and the cooperative staff who ran the utility. To some extent, this relationship has been preserved. Despite the recent trend away from acceptance of the "cooperative philosophy," the majority of cooperatives have worked to preserve the sense of cooperation and shared goals that define that philosophy.

The fact that the consumer owns shares in the cooperative and is regularly invited to attend meetings at which his peers discuss business related to his utility results in an attitude different from that of the typical utility customer. This attitude can be observed in the different customer acceptance rates for incentives designed to encourage direct load control. According to a 1985 survey,[1] 2,530,174 consumers participate nationally in direct load control (an increase of 69% over 1983). Twenty-three percent of those participating are cooperative consumers, yet only 65% of the cooperatives involved in these programs offer incentives to their members to encourage participation, compared to 94% for investor-owned utilities. Cooperative consumers seem more likely to participate in DSM programs because, in effect, they have a personal stake in the success of such programs.

To the extent that DSM programs can be used to balance supply/demand relationships in the short term (e.g., up to three years), they have tremendous potential value to the cooperative and its members in addition to the longer-run benefits previously described. These benefits, however, depend critically on the successful integration of the activities of a distribution cooperative with the generation and transmission (G&T) cooperative supplying it, and with the other distribution cooperatives purchasing from the same G&T. Clearly, a generation and transmission cooperative that serves distribution cooperatives who are pursuing conflicting load shape objectives will not be well positioned to benefit from the planning flexibility (relative to supply-side alternatives) inherent in DSM programs.

SELECTING THE RIGHT DSM ALTERNATIVE

DSM encompasses the planning, evaluation, implementation, and monitoring of activities selected from a wide variety of technology and marketing alternatives. This wide range of alternatives mandates that RECs should seriously consider including DSM in their overall planning process.

However, assessing which alternative to use involves careful examination of the cooperative's purchased power arrangements, expected load growth, capacity expansion plans, load shapes, etc. Because of the variance in these factors, it is often inappropriate to transfer DSM programs from one service area to another without allowing for necessary adjustments. In addition, the success of any alternative depends on the specific combination of marketing activities selected to promote consumer acceptance.

DEMAND-SIDE MANAGEMENT IN THE CONTEXT OF UTILITY PLANNING

In examining the traditional utility planning framework shown in Figure 12-1, it is apparent that electric utilities historically have accepted the level of demand for their product, as well as the hourly and seasonal load shapes, as a fixed input (Box A) and proceeded to build or acquire the least-cost generation mix (Box B) to meet that load. Given the generation mix in place, production costing methods (Box C) determined how to dispatch the system to meet the load at minimum operating cost. Since revenue requirements have been largely defined by such engineering/economic models, decisions left to cooperative management have been essentially limited to those relating to financing of the system and allocation of costs among consumers.

DSM introduces a new element into the planning framework by allowing for deliberate change in the load shape in order to pursue strategic objectives. Such changes may, in turn, change generation requirements and mix, and generator dispatch, therefore affecting the cooperative's costs and financial requirements. Thought of in these terms, DSM clearly can combine with traditional supply-side alternatives in a resource planning portfolio that can greatly increase the flexibility and manageability of an electric cooperative. As previously mentioned, the supply/demand integration implied by this approach to planning requires a highly coordinated effort between distribution and generation/transmission cooperatives.

Figure 12-1. TRADITIONAL UTILITY PLANNING FRAMEWORK

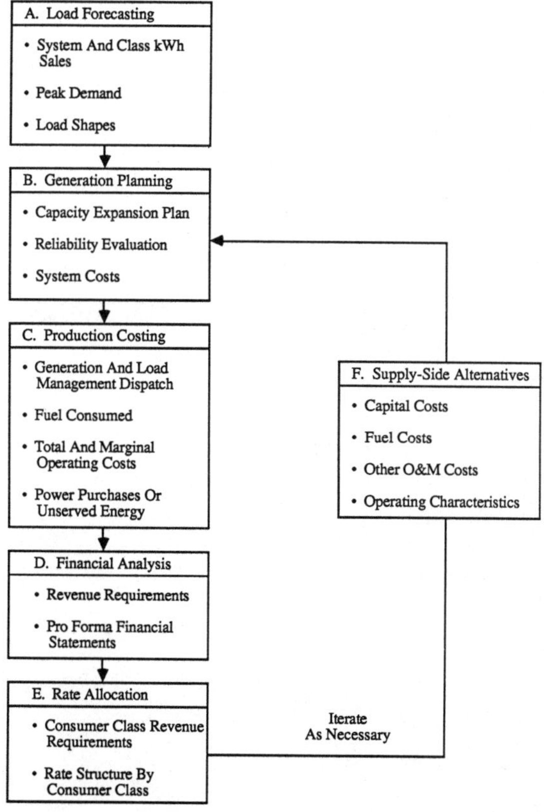

Source: Battelle-Columbus Division[2]

An example of how demand-side management can be integrated within an REC's planning process is shown in Figure 12-2. This figure illustrates the demand-side planning framework developed by Synergic Resources Corporation for the Virginia, Maryland and Delaware Association of Electric Cooperatives, Old Dominion Electric Cooperative (G&T), and their member cooperatives.[3] The process ensures that two general divisions of planning activity are undertaken. The G&T is responsible for traditional supply-side resource planning and for identifying and determining the technical feasibility of demand-side options. The latter step includes analysis of the load shape impacts of programs as well as evaluation of supply- and demand-side strategies based on least-cost planning. The Association's role is to identify and screen demand-side strategy options, work with member cooperatives in market research, and assist in the training and implementation of demand-side programs. The member cooperatives' role is to identify and recommend supply- and demand-side strategies, supply inputs to benefit/cost calculations, and implement and evaluate the programs.

MARKET IMPLEMENTATION METHODS

A key determinant of the success of DSM alternatives is the selection of the appropriate market implementation methods. Managers of cooperatives can choose from a wide range of methods designed to influence consumer adoption, which can be broadly classified into six categories:

- Consumer education
- Direct consumer contact
- Trade ally cooperation
- Advertising and promotion
- Alternative pricing and rate structures

Figure 12-2. DEMAND-SIDE MANAGEMENT PLANNING PROCESS

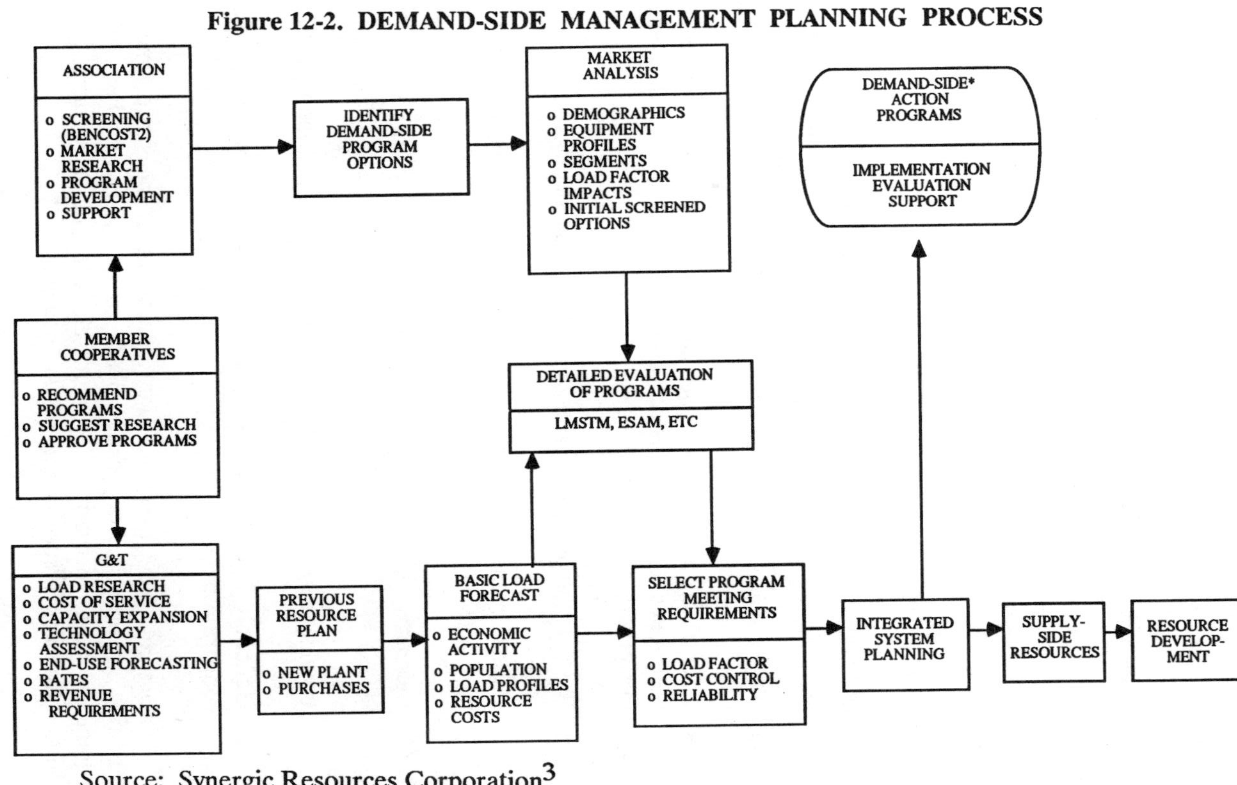

Source: Synergic Resources Corporation[3]

- Direct incentives.

The primary purpose of each of these market implementation methods is to influence the marketplace and to change consumer behavior. Cooperative managers must choose the method(s) that will obtain the desired consumer acceptance and response. A brief description of each method, as it applies to cooperatives, is provided below.

Consumer Education

Most cooperatives rely on member education programs to promote general member awareness of DSM programs. Such education is the most basic of the market implementation methods available to a cooperative, and can be used to:

- Inform members about products/services being offered and their benefits, and influence consumer decisions to participate in a program

- Increase the perceived value of service to the member

- Inform members of the eligibility requirements for program participation

- Increase the member's knowledge of factors influencing energy purchase decisions

- Provide members with other information of general interest

- Generally improve relations between members and cooperatives.

Whenever a new program is introduced or changes are made in an existing program, a cooperative needs to notify its members. If a program is new, increasing the general level of awareness is an important first step in encouraging market response.

Education is the most widely applicable market implementation method for DSM measures. Some cooperatives' member education brochures describe the operation, benefits, and costs of DSM technologies (e.g., heat pumps and thermal storage). Others focus on methods to reduce energy costs. Typically, utilities also provide do-it-yourself guides on home energy audits, home weatherization, and meter reading, as well as information on DSM programs and member eligibility.

The major advantage of member education techniques is that they typically provide a more subtle form of marketing, with potential to positively influence member attitudes and purchase decisions. Member education techniques should be used in conjunction with one or more of the other market implementation methods for maximum effectiveness.

Direct Consumer Contact

Direct consumer contact techniques refer to face-to-face communication between the consumer and a cooperative representative to encourage greater consumer acceptance of DSM programs. Many cooperatives have employed marketing and consumer service representatives to provide advice on appliance choice and operation, sizing of heating/cooling systems, lighting design, and even home economics.

Examples of direct consumer contact programs include energy audits, program services, storefront information centers, workshops/energy clinics, exhibits/displays, and quality/safety inspection programs. The Northern Virginia Electric Cooperative's Earth Coupled Heat Pump Program makes liberal use of direct member contact as well as member education. The program is targeted toward builders, heating/cooling contractors, and custom home buyers in order to gain their awareness, interest, and acceptance of the earth coupled heat pump. Initially, earth coupled heat pumps will be targeted toward higher priced, upscale homes. The following features are included in the program:

- A workshop on earth coupled heat pumps
- Open houses featuring homes using earth coupled heat pumps
- A builder and heating contractor handbook on earth coupled heat pumps
- A member brochure on earth coupled heat pumps
- Sizing and cost analysis of earth coupled systems compared to other heating/cooling systems
- A contest for trade allies providing a $1,000 award for the highest number of earth coupled heat pump installations in the service territory.

Homeowners who install the earth coupled heat pump will be recognized with plaques identifying their homes as having "high tech" comfort systems. Northern Virginia will also meter at least five of the homes in order to monitor the performance of the systems.

Direct consumer contact methods are applicable to a wide range of DSM options. A major advantage of these methods is that they allow a cooperative to obtain feedback from the consumer, thus providing an opportunity to identify and respond to major consumer concerns. They also enable the cooperative to perform more personalized marketing, and can be useful in communicating the cooperative's interest in and concern for controlling consumer energy costs.

However, direct consumer contact methods are labor-intensive and may require a significant commitment of cooperative staff and other resources. Also, specialized training of cooperative personnel may be required. Other issues that a cooperative may face relative to direct consumer contact include:

- Scheduling requirements, as well as the need to respond to consumer concerns in a timely and effective manner

- Potential constraints arising from the need for field service personnel to implement direct consumer contact programs in addition to their other duties

- Possible opposition by local contractors/installers to direct installation programs

- Liability issues related to inspection and installation programs

- "Fair trade" concerns related to contractor certification for audits or installation.

Trade Ally Cooperation

Trade ally groups can be extremely useful to cooperatives in promoting a variety of DSM measures and can also help reduce the cooperative's implementation costs. In addition, they can provide technical, logistical, and consumer response information that is useful in the design of cooperative programs. Generally, if trade ally groups believe that a certain cooperative program will help them, they are likely to support it.

An example of such a program is Southern Maryland Electric Cooperative's Watt-Wise Central Air Conditioner and Heat Pump Program. The program involves heating and cooling contractors in a cooperative marketing effort. The following features are included in this program:

- Heating, ventilation, and air conditioning (HVAC) contractors receive cooperative advertising and referrals.

- Southern Maryland Electric Cooperative, with assistance from contractors, provides members and builders with system sizing and equipment cost analysis.

- A sweepstakes is offered presenting a $1,000 award to the contractor who installs the greatest number of high efficiency air conditioners and heat pumps.

An important part of the program involves promoting the addition of home energy conservation services so that the size and run time of air conditioners can be reduced. Southern Maryland has also considered adding financing provisions for conservation and replacement cooling equipment by using REA Bulletin 20-23; loans of up to $3,000 could be offered for each single family dwelling in cooperation with local banks.

To obtain the greatest benefits from trade allies, cooperatives must be willing to compromise and accommodate allies' concerns and questions related to product availability, certification requirements, reimbursement of expenses, cooperative advertising/promotion, training needs, etc. One major benefit of such cooperation is that DSM programs are less likely to face legal challenges related to fair trade if these programs are cosponsored with trade allies.

Advertising and Promotion

Cooperatives have used a variety of advertising and promotional techniques to communicate informational and persuasive messages to consumers. Advertising media applicable to DSM programs include radio, television, magazines, newspapers, outdoor advertising, and point-of-purchase displays.

Promotion usually includes activities that support paid advertising, such as press releases, personal selling, displays, demonstrations, coupons, and contests/awards. Some cooperatives prefer to use newspapers, based on consumer awareness of DSM programs. Others have found television or radio advertising to be more effective.

Like consumer education methods, advertising and promotion have widespread applicability. A number of innovative radio and television spots, highlighted by appealing slogans, jingles, and humorous dialogue, have been developed by electric utilities to promote DSM measures. Other promotional techniques used by electric utilities have included awards, energy efficient home logos, and residential home energy rating systems.

Some media may be more useful than others in terms of moving consumers from the "awareness" stage to the "adoption" stage in purchasing a product or service. Therefore, a multifaceted and carefully scheduled advertising/promotional campaign is worthy of consideration as part of any rural electric cooperative's DSM plan.

Alternative Pricing and Rate Structures

Alternative pricing, through innovative rate structure changes, can be an important implementation technique for cooperatives promoting DSM options. Rate incentives for encouraging specific patterns of utilization of electricity can often be combined with other strategies (e.g., direct incentives) to achieve a cooperative's DSM goals. A major advantage of alternative pricing programs over some other types of implementation techniques is that the cooperative has little or no cash outlay. The member receives a financial incentive over a period of years, so the cooperative can provide the incentives as it receives the benefits.

One potential disadvantage to some alternative pricing schemes, such as time-of-use and rates involving demand charges, is the cost of metering, which can sometimes amount to several hundred dollars per installation. In addition, such rates do not lower the consumer's purchase cost, as do some other implementation techniques such as rebates. Note also that detailed information regarding cost of service and load shapes is needed to design and implement effective rates that serve to promote DSM options. It may be necessary to educate members regarding rate structures and related terminology.

Member response to and acceptance of alternative rate structures will vary, depending on the characteristics of the member, the DSM programs, and the specific rate structures. Research done by the Electric Power Research Institute (EPRI) on consumer response to time-of-use rates indicates that:

- On average, considerable similarity exists in consumer response patterns across several geographically diverse areas.

- Households typically reduce the share of peak usage in weekday summer use by three to nine percentage points.

- The degree of consumer response is influenced by appliance ownership and climatic conditions.

Direct Incentives

Direct incentives are being used in a large number of rural electric cooperative DSM programs to encourage consumer participation. Various types of direct incentives are applicable to many of the cost control/consumer options in each of the major option categories.

An example of a direct incentive program is Mecklenburg Electric Cooperative's Residential Energy Efficiency Loan Program. The program is offered in conjunction with Rural Electrification Administration (REA) financing assistance. It provides subsidized loans to finance the installation of energy efficiency measures in residential owner-occupied homes. The minimum loan available is $500; the maximum, $3,000. All loans are for a period of six years, with APR financing set at one-half the current consumer interest rate. The loans are not available to households with incomes over $50,000/year.

The primary marketing technique used to promote this program is a bill insert; an insert is also placed in Rural Living Magazine, a monthly publication received by Mecklenburg's members. Members interested in the program return a reply card attached to the insert. The reply card includes questions to be filled out by the member concerning information on housing characteristics (i.e., insulation level, age and type of heating system).

Members who are deemed eligible for the program receive a walk-through audit by a Mecklenburg staff member. The measures needed and their estimated costs are determined. It is then up to the homeowner to contact one of a number of cooperative-approved contractors to perform the work; the cooperative provides assistance in finding a reputable contractor. The contractor bills the consumer directly. A check is issued to the consumer from the cooperative; both the consumer's and the contractor's names are on the check. The consumer gives the check to the contractor once there is a signed agreement that the work has been done satisfactorily. The cooperative then sets up a payment schedule with the member in addition to the regular monthly bill.

In developing direct incentive programs, cooperatives should be aware of potential fair trade or antitrust concerns. However, as long as a cooperative establishes rational criteria for dealing with trade allies and uses sound practices, it is unlikely that the legality of incentives would be challenged.

Both consumer acceptance of and response to direct incentive programs, and the potential costs and benefits to the cooperative that result from such programs, depend on the nature of the incentive, the technology being promoted, and the consumer's socio-demographic and economic characteristics. Utility experience with these programs indicates that participants are generally middle to upper class households.

CASE STUDIES

Following are three detailed case studies of cooperative DSM programs. The specific approaches taken by these cooperatives to address what they perceived as their strategic objectives can be instructive to other planners, even if these planners have very different objectives.

Buckeye Power, Inc.

Buckeye Power, Inc. is a generation and transmission cooperative that provides all the wholesale electric power requirements of the 29 rural electric cooperatives operating in the State of Ohio. Buckeye's DSM program had its origins in the early 1970s. At that time, Buckeye became very concerned with its escalating peak demands and deteriorating annual load factor and determined that something must be done to reverse such trends. The rates facing rural electric cooperative consumers were generally higher than those being paid by the consumers of Ohio's investor-owned utilities. Thus, the ultimate objective of Buckeye's DSM program was to stabilize or reduce the cost of wholesale electricity to its member cooperatives. The tactical objective resulting from this overall strategic objective of Buckeye was to delay the time at which the utility would have to add new, "expensive" generating capacity.

Buckeye personnel reviewed several possible alternatives for DSM, but settled on a strategy to control water heaters. This decision was reached because water heaters represent the second highest demand appliance and because their heat storage properties meant that they could be controlled with little or no inconvenience to their consumers. Drawing on results from the Pioneer Rural Electric Cooperative in Piqua, Ohio, in which radio control of 150 electric water heaters was demonstrated, a Buckeye system-wide feasibility study was completed. This study, completed in 1972, indicated that the implementation of a water heater control program would be cost-effective if each control point would yield as much as 500 watts of diversified load reduction during the time of Buckeye's system peak. Since the estimate of diversified load reduction was 1.2 to 1.4 kW per control point, the decision was made to implement a voluntary program of radio control of water heaters.

At last report, there were approximately 53,000 radio switches installed on electric water heaters in the Buckeye member cooperative service territories. During the system peak in December 1983, about 1 kW of peak load reduction per radio switch was attained. This reduction of 53 MW of peak load was quite beneficial, since Buckeye's costs average about $86,000 per MW of peak load, which is ratcheted on a 100% basis. The savings from this system can amount to an excess of $4 million per year. Since the annual cost of owning and operating the water heater control system amounts to about $1 million, system implementation results in a net annual savings of about $3 million dollars.

Oglethorpe Power Corporation

Oglethorpe Power Corporation is the generation and transmission (G&T) cooperative providing wholesale electric power to 39 Electric Membership Corporations (EMCs) in Georgia. The decision to implement DSM programs rests with the individual EMC Boards of Directors. Since the EMCs are eligible for 5% financing from REA for 70% of their capital investment requirements, moving DSM implementation responsibility to the EMCs can lower power costs while increasing local EMC involvement. Typically, an EMC has a weighted cost of capital below 8%, whereas Oglethorpe would face a cost of capital closer to 11%. Recognizing the high cost of constructing new generation facilities and the differences in financing costs, Oglethorpe has developed a market basket of DSM alternatives for consideration by each of its 39 members. Oglethorpe is currently developing a spreadsheet handbook to help the EMCs choose among the alternatives that have been identified as being potentially cost-effective for the system as a whole.

The EMCs' overriding objective in implementing DSM programs is to reduce power costs for the EMC systems and ultimately for the retail consumer. Oglethorpe generates more than one-half of the electric energy requirements of its members and purchases the remaining capacity and energy from Georgia Power Company. Capacity purchases are currently based on the contribution of the EMCs to the highest 44 hours of peak demand experienced annually on the Georgia Territorial System. Therefore, many of the DSM options that Oglethorpe has analyzed and is promoting are oriented toward reducing Oglethorpe's system demand during peak hours for the Georgia Territorial System.

Not all DSM options are new to Oglethorpe's EMCs. Over the past five years, Oglethorpe has coordinated the development of a statewide program for installation and operation of direct control switches. Approximately 200 MW of demand reduction has been achieved by Oglethorpe's EMCs through a combination of direct control of air conditioners, water heaters, and irrigation pumps, and voltage regulators for system voltage reduction. Although 28 of Oglethorpe's 39 EMCs are currently involved, there is a significant potential for expansion of this statewide program to achieve further peak demand reduction.

Oglethorpe's analysis estimates that peak demand can be reduced an additional 140 MW by 1994 with the installation of additional air conditioner controls. This estimate is based on an assumption of adding controls to 12,700 new households enrolling in the program each year between 1985 and 1994 with a demand reduction potential of approximately 1.1 kW per switch. Oglethorpe estimates that an additional 140 MW of peak load reduction can be achieved by 1994 through direct control of additional water heaters. This estimate is based on a projection of approximately 28,000 installations of water heater switches per year, with approximately 0.5 kW peak reduction per switch.

The third DSM program being pursued by Oglethorpe Power is the direct control of irrigation pumps. Over the ten-year time frame from 1985 through 1994, there exists a potential to install direct control switches on approximately 189 pumps per year at an average savings of 32 kW per pump. These controls would be placed on irrigation systems for Georgia crops which would include soybeans, peanuts, and tobacco.

Another direct control program evaluated by Oglethorpe is a combination of voltage reduction during peak hours and system-wide power factor improvement. It is expected that voltage reduction by direct control can reduce peak demand by approximately 50 MW by 1994. This would be achieved by reducing distribution voltage by four volts during selected peak hours on selected distribution feeders.

By implementing a program of time-of-use rates for approximately 50% of the residential consumers served by Oglethorpe's EMCs, Oglethorpe could expect a reduction in peak demand of approximately 450 MW in 1994. This level of load reduction is based on estimates of response to a very simple time-of-use energy rate. This rate would have a peak to off-peak energy price ratio of four to one. The on-peak period would be defined as the times between noon and eight p.m. on weekdays during the months of June through September. The off-peak rates would be in effect during the nights and weekends of summer and all other times from October through May.

Oglethorpe is also promoting insulation programs in an effort to reduce peak demand. Oglethorpe estimates that its home insulation program can reduce peak demand 240 MW in 1994 by increasing ceiling insulation to the equivalent of 12-inch fiberglass, and by weather stripping doors and windows in 60% of the eligible households by 1994. The analysis of this program assumed that each EMC would offer a four-year interest free loan to the consumer installing this insulation.

Oglethorpe is also encouraging its EMCs to adopt a water heater insulation program. This program has the potential for reducing forecast peak demand by approximately 17 MW by 1994. In this program a four-inch fiberglass insulation wrap or its equivalent would be installed on 75% of the electric water heaters.

In Georgia, large consumers with at least 900 kW of connected load have the option of choosing their power supplier at the time they first request electric service. With new industry locating in Georgia and increasing large commercial development, it is not unusual for one of Oglethorpe's EMCs to compete with other Georgia electric utilities to provide service to a new establishment. The local EMC must consider a number of factors in deciding how vigorously it will compete for these new loads; one factor it considers is the impact of the new load on power costs.

Oglethorpe has analyzed the impact of the addition by the EMCs of 200 MW of new industrial load between 1985 and 1994. This analysis has shown that adding industrial loads with a billing load factor of at least 90% can lower future power costs. For example, if 200 MW of industrial load were added with a billing load factor of 100%, the average EMC would see a long-term savings in power costs of about $2 million, and the savings for Oglethorpe would total over $45 million. Oglethorpe has also encouraged its EMCs to adopt programs that encourage consumers to install electric heat pumps in new and existing houses.

Minnkota Power

Minnkota Power Cooperative is a generation and transmission cooperative that provides wholesale power for 12 rural electric cooperatives in the eastern part of North Dakota and the northwestern part of Minnesota.

Minnkota's DSM program began in 1973. The cooperative faced a situation in which the winter peak was growing rapidly due to the market penetration of electric space heating. The rapidly rising costs of oil and propane, combined with the availability of relatively low-priced electricity from Minnkota's distributors, had caused Minnkota's winter peak loads to grow disproportionately. Further, because of the climate in areas served by Minnkota and its 12 distribution cooperatives (the area has an average of 9,500 heating degree days annually and 450 cooling degree days), there was relatively little potential for air conditioning loads in the summer to offset the growth in the winter peak load.

The major DSM effort that has been undertaken by the Minnkota Power system is a dual-fuel heating program. This concept is one in which the end user has a dual furnace which uses electricity for the most part, but which is switched to oil during hours of system peak. This allows Minnkota to have the electricity sales associated with space heating without having negative impacts on annual load factors. The dual-fuel program was the preferred alternative for Minnkota because electric heat represented the largest component of growth in required capacity and the highest cost component of the consumer's utility bill. Therefore, it seemed to offer the greatest incentive for application of DSM from both Minnkota's and the consumer's points of view.

After thorough research of radio control, carrier, and ripple control systems during 1974 and 1975, Minnkota opted for a ripple control system. Radio control was unattractive to Minnkota at the time because of its limited operating flexibility and the need for 30 transmitter sites to cover the service area. Carrier control would have required over 130 control site installations, and at the time, was not considered to have a proven track record.

The ripple control system provided needed coverage with 12 injection sites. This decision to go with ripple control was made on the basis of "attractive economics, ready availability, proven reliability, flexibility in application, and the unique requirements of our intended application" and was subsequently funded by the Rural Electrification Administration.

Beginning in 1975, Minnkota initiated its dual-fuel market program. Intuitive judgment was that 50 to 70% market penetration for the dual-fuel program could be expected if electricity was priced low enough to make it economically attractive in comparison to the principal alternative (i.e., heating by oil or propane). Currently Minnkota has a total of about 19,000 dual-fuel heating systems installed with a total interruptible capability during peak conditions of 190 MW. This market penetration was accomplished through rate design, rebates to consumers installing the dual-fuel systems, and media promotion of the concept. Minnkota's wholesale rates to the distribution cooperatives are $92/kW year for the base demand, which is equal to the member's average demand for the 1973-1974-1975 years plus 10%; demand in excess of this is billed at $184/kW year. This rate structure provides a strong incentive for the distribution cooperatives to keep their peak period demand down, and most of them have passed this price signal to consumers through special interruptible rates (2.25 to 3 cents/kWh) for those involved in the dual-fuel program.

In addition, the dual-fuel program has been promoted through information sharing, advertising, and through a rebate in which Minnkota and the distribution cooperative provided 50% of the necessary capital costs required to install a dual-fuel system. Many of the distribution systems do their own advertising and promotion; there is a wide variance in the type and degree of promotion that is undertaken. No analysis has yet been conducted comparing the level of market penetration for different types of promotional campaigns.

In addition to these residential dual-fuel units, two commercial dual heating installations -- one 30 MW and the other 40 MW -- give Minnkota a total interruptible capability of 220 MW. Since total firm load is generally 380 MW during the winter season, the system can operate anywhere between 380 and 600 MW during a given winter season.

In addition to the dual-fuel program, Minnkota has several other DSM programs in place. Loads which may be shed for periods of up to eight hours include water heaters and potato warehouses. Currently, Minnkota has 6,000 kW of interruptible electric heating of potato warehouses, which can be shed for a period of four hours by use of time clocks. A future expectation is that water heaters will be controlled in buildings where ripple receivers are installed to control other more important loads such as slab storage heat or dual heat.

A second category of controlled load that is a major part of Minnkota's DSM program is comprised of those loads that may be shed for periods of 8 to 16 hours. These residential loads consist primarily of slab storage heating in which electric cables are buried in sand below the basement floor slab. The sand and slab are used as a heat reservoir, which is charged for 12 hours during off-peak periods and discharges its thermal energy to the area above the slab during peak periods. Currently, Minnkota has about 5,000 kW of load controlled by ripple receivers in this manner.

There have been two major results of the Minnkota DSM programs. First, reduced heating energy costs of 30 to 50% and an enhanced communication program have generated confidence and goodwill among consumers, which is considered a very significant achievement; secondly, the annual load factor has improved from a low of 48% in 1976 to 63% in 1983, with a goal of 70% by 1985.

CONCLUSION

America's rural electric cooperatives have become increasingly involved in the implementation of demand-side marketing programs in an effort to promote the wise and efficient use of electric energy. Demand-side programs can be integrated into the REC's planning framework and used to combat the many obstacles which threaten the competitiveness of their electric service. Cooperatives have recognized the value of various demand-side options and, with the help of their generation and transmission cooperatives, have begun to design innovative programs which match individual requirements. These requirements are based on member needs -- a vital strategy for any organization's success.

REFERENCES

[1]Plexus Research Inc., 1985 Survey of Utility Residential End-Use Projects, prepared for Electric Power Research Institute, EM-4578, May 1986, pp. 1-28 through 1-30.

[2]Battelle-Columbus Division, Demand-Side Management for Rural Electric Systems, prepared for Electric Power Research Institute and National Rural Electric Cooperative Association, 1986.

[3]Synergic Resources Corporation, Demand-Side Management Implementation Considerations for the Association and Old Dominion, prepared for The Virginia, Maryland & Delaware Association of Electric Cooperatives, August 1986.

[4]Battelle-Columbus Division and Synergic Resources Corporation, Demand-Side Management - Volumes 1-3, prepared for Electric Power Research Institute and Edison Electric Institute, EA/EM-3597, August 1984.

CHAPTER 13

Marketing Residential Heat Pumps

Clark W. Gellings

INTRODUCTION

Approximately 45 utilities in the United States have heat pump marketing programs. These programs include customer education and advertising programs, direct contact programs, financial incentives, trade ally programs, and alternative rates. Some utilities are beginning to "bundle" two or more marketing techniques together in order to add greater value to customers. For example, a heat pump marketing program that not only positions the heat pump to customers in a meaningful way but which also provides economic benefits will be important. Programs involving trade allies are also becoming more important because of the critical input they provide in evaluating heating/cooling system options.

At first glance, it appears that there is a wide range of success in heat pump marketing programs. However, even after the obvious differences among service territories -- such as competition, relative prices, and popularity of heat pumps in a particular area -- have been removed, it still appears that there is little consistency in the relative effects of rebates, alternative rates, and promotion. As an industry, our understanding of customer preference and behavior and our ability to mold a "heat pump preference" in a customer's mind has barely begun. As we begin to examine the customer's view of electricity purchase/use/consumption, the puzzle of which marketing programs work best in each location begins to fall in place.

THE CUSTOMER'S PERSPECTIVE

In order to understand the utility customer's view of electricity purchase/use/consumption, it is helpful to examine a general model of consumer decision-making behavior. Several of the preceding chapters in this book -- particularly Chapters 2, 3, 4, and 5 -- focus in detail on this model. As a brief recap, however, here are the four stages of the consumer decision-making process:

- Problem recognition (need recognition/awareness)
- Alternative search and evaluation (search for alternatives/interest)
- Actions (purchase/adoption)
- Post-action behavior (satisfaction).

TYPES OF MARKETING

As described in more detail in Chapter 12, electric utility market implementation methods can best be grouped into the following six categories:

- Customer education
- Direct customer contact
- Trade ally cooperation
- Advertising/promotion
- Alternative pricing
- Direct incentives.

Each of these methods can be used in a heat pump marketing program to influence various stages of the customer decision-making process. Figure 13-1 depicts the general applicability of these methods for removing barriers toward buyer acceptance.

Figure 13-1. APPLICABILITY OF MARKET IMPLEMENTATION METHODS TO STAGES OF BUYER READINESS

	Market Implementation Method					
Stage of Buyer Readiness	*Customer Education*	*Direct Customer Contact*	*Trade Ally Cooperation*	*Advertising/ Promotion*	*Alternative Pricing*	*Direct Incentives*
Need Recognition/ Awareness	●	○	○	○		
Search for Alternatives/ Interest	○	●	●	●		
Purchase/ Adoption		○	●	○	●	●
Satisfaction		○	○			

●*Generally applicable* ○*Conditionally applicable* *[BLANK] = Rarely or not applicable*

Source: Battelle-Columbus and Synergic Resources Corporation[1]

Utility heat pump marketing programs should consider all these methods in addressing the issues surrounding the heat pump purchasing decision for a particular customer segment. The marketing program may consist of one of the market implementation alternatives or several used in combination; the goal is to influence the customer at one or more key points in his decision-making process. Table 13-1 indicates the applicability of each market implementation method to the state of buyer readiness. Successful heat pump programs usually utilize several market implementation methods so as to address each stage of buyer readiness.

Table 13-1. HOW MARKET IMPLEMENTATION METHODS CAN AFFECT THE STAGES OF HEAT PUMP BUYER READINESS

Stage of Buyer Readiness	Customer Education	Direct Customer Contact	Trade Ally Cooperation	Advertising/ Promotion	Alternative Pricing	Direct Incentives
Need Recognition/ Awareness	Build awareness and interest Communicate benefits and features	Build awareness and interest	Build awareness and interest	Build awareness and interest Communicate benefits and features Identify utility programs Identify trade allies		
Search for Alternatives/ Interest		Properly estimate costs	Third party advice	Maintain interest Identify heat pump		
Purchase/ Adoption		Properly size Close the sale	Distribution channel Close the sale	Close the sale	Improve life-cycle cost	Improve life-cycle cost
Satisfaction		Inspect installation and performance Service and maintain	Service			

Collectively, then, these elements constitute a marketing program. Programs are designed and targeted toward certain market segments. Not all utility customers are alike, and the differences among them can be reflected in widely varying acceptance rates for heat pump technology. This fact suggests that utilities should attempt to segment their heating/cooling markets by using demographics, location, psychographics, and usage level rates in order to focus their marketing strategies.

The electric utility industry needs to gain a much more sophisticated understanding of customer acceptance of and preferences for various heating and cooling systems. The industry needs to optimize marketing program bundles in order to gain the acceptance levels required to make marketing programs cost-effective, and understanding customer behavior is critical to the effective design of these program bundles.

Segmenting the heat pump market according to multiple criteria can be a very helpful way to focus on the customer group that is most likely to respond to a heat pump marketing program. Segmenting the market can also provide information on what, where, and how to best communicate with the various customer segments. There is clearly a segment of the population that desires full comfort, is prone to try new technologies, is looking for cost control, and desires low-risk technology. This is the ideal segment to which heat pump technology can be marketed.

The Electric Power Research Institute (EPRI) initiated a major marketing project in 1986 to identify this important customer segment of heat pump purchasers.[2] This project involves the use of focus groups, in-depth interviews, a national survey, and modeling/simulation of heat pump purchases. Test marketing may also be an important component of this research. After this research is complete, the industry will be able to more consistently develop marketing programs that are reliable and efficient.

Suggestions on how to segment the market can be obtained from reviewing results of an ongoing EPRI project on demand-side management.[3] In this project, customer acceptance of DSM programs is being analyzed. Preliminary findings indicate that the reasons why customers participate in heat pump programs include the following:

- Customers want to reduce the cost of heating their homes and they want air conditioning as a "side benefit."
- Customers' supply of fuel oil is uncertain.
- Customers like third party inspection of the installation.
- Customers like the financial incentives.
- Customers desire to become more energy efficient.
- Customers are in need of a new space-conditioning system.

Reasons why customers do not participate in heat pump programs include the following:

- The old system is working fine.
- Heat pumps cost too much to install.
- To install a heat pump would require that the entire system be changed out.
- Customers are skeptical of heat pump performance and the somewhat higher initial costs.
- End-use customers do not fully understand the technical operation of a heat pump.
- They distrust heat pumps.

CONSUMER DECISION-MAKING PROCESS

Table 13-2 illustrates the market structure for residential heating, ventilation, and air conditioning. The structure is comprised of markets, market type, equipment purchasers, those having key market influence, and installers/retailers/servicers. A combination of markets and market types yields market structure. To examine the steps in the consumer decision-making process in the context of the purchase of a heat pump, it is helpful to use five market segments (types A through E) representing single and multifamily dwellings (excluding mobile homes) as follows:

Table 13-2. RESIDENTIAL HVAC MARKET STRUCTURE

Markets	Market Type	Equipment Purchaser
Single Family	New Construction	Owner/Occupant
Multiple Family	Major Remodeling	Owner/Leaser
Mobile Home	Replacement	Builder
	Retrofit	Developer
	Mobile Home Manufacturer	Mobile Home Manufacturer
	Mobile Home Accessory Dealer	

Key Market Influence	Installers/Retailers/Servicers
Trade Allies	Mass Merchandisers
Utilities	Appliance Dealers
Government	Electrical Contractor
Oil Dealers	Plumbing Contractor
	Air Conditioning Contractor
	Mechanical Contractor
	Mobile Home Manufacturer
	Mobile Home Dealer

Source: Electric Power Research Institute[4]

- Type A -- New home speculative builder
- Type B -- Residential consumer building new home

- Type C -- Residential consumer about to purchase add-on central air conditioning

- Type D -- Residential consumer whose heating system has failed

- Type E -- Residential consumer who has existing, operating fossil heating system

Recalling the four-step decision-making process -- problem recognition, alternative search and evaluation, actions, post-action behavior -- the following section examines each of these decision-making steps as they relate to the purchase of heat pumps.

Problem Recognition

This stage is rather straightforward for types A through D. In each case, the need to select a heating system is obvious. The type E segment, however, may not reach the stage of recognizing that there is a problem without a little help. Some type E customers are satisfied with the cost and other attributes of their fossil-fueled heating systems. Other customers in this segment may be aware of their dissatisfaction with the high cost of gas or oil heating for their home, but they may not have realized that the problem is related to their heating system or fuel. On the other hand, these customers may be fully aware of the reasons for their dissatisfaction and thus ready to begin the alternative search.

Those who are satisfied with their heating, and those who are dissatisfied but have not begun a focused search, represent an opportunity for the utility. By including elements in its marketing program that will alert these customers to possible dissatisfaction and focus them toward a search which includes the heat pump as an option, the utility may be able to penetrate a market segment which at first glance seems closed.

For utilities just starting heat pump promotional programs, it is extremely important to begin by building awareness and positioning the technology in ways meaningful to the end user. Figure 13-2 suggests that there is very low awareness (only 15%) of heat pump technology in some utility market areas. Utilities promoting heat pumps should strive to increase awareness levels to at least 65% in the target market.

Alternative Search and Evaluation

In this stage, the behavior of the five segment types is more varied. Type A is unique in that he is concerned solely about his own profits. He will buy the lowest capital cost system that will "sell" the home quickly at an optimum price. In this stage, the builder needs to be convinced that the higher capital cost will make his homes more marketable. Alternatively, direct incentives which lower the capital cost will persuade the builder as long as he doesn't perceive any negative reaction to heat pumps in the real estate marketplace.

Figure 13-2. KNOWLEDGE OF HEAT PUMP OPERATION

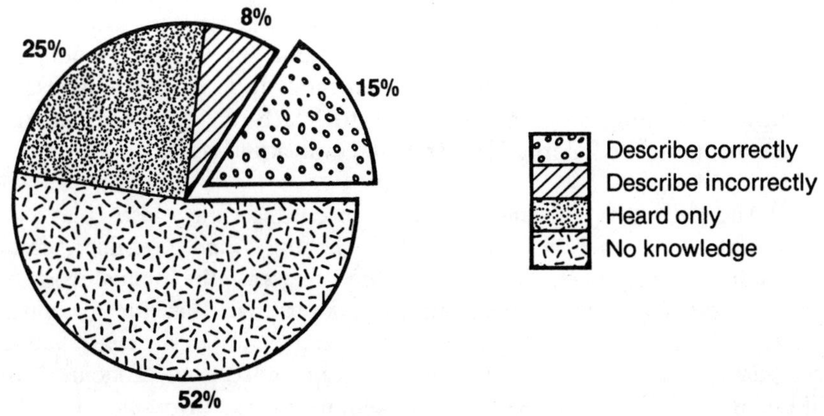

Source: Davis[5]

The type B segment needs to be made aware of the availability of a hcat pump and its features. The type C segment needs to be informed that an add-on air conditioner could offer heating, too. Use of trade allies is crucial in addressing this market segment, as it is for the type D segment. Once alerted to alternatives, type E will behave in a similar fashion. The utilities' role in types C and E may involve direct contact to ensure that the heat pump is adequately analyzed. Types B through E are all interested in "life-cycle costs," though not all individuals within each type have the same discount rates (or even use that terminology).

From a technical standpoint, there are a number of benefits that a heat pump offers utility customers. First, the heat pump delivers more energy than it consumes. Electric heat pumps have efficiency levels greater than 200%. Typical air-source heat pump models have a seasonal performance factor (SPF) of 2.0, going as high as 2.6. Ground-coupled heat pumps can provide SPFs greater than 3.0. (The actual SPFs of heat pumps vary by region and seasonal use.) The superior efficiency of the heat pump means less energy is needed to meet the heating and cooling needs of customers. In addition, heat pumps can be downsized and their operating costs further reduced if they are installed in more efficiently designed dwellings. Some utilities have been promoting heat pumps in new, efficiently designed homes and have been able to reduce the size of the heat pump significantly. The significant benefit offered by a heat pump is comfort at a low cost. This fundamental fact must be stressed through marketing programs designed to reach the targeted market segments during the "alternative search and evaluation" stage.

Alternative rates and financial incentives may also play an important role in making electric heating and cooling costs competitive. For example, in some service areas, the annual cost of using a gas forced-air furnace may be lower than that for a central electric heat pump. However, for some utilities, conservation programs and alternative rate designs can make the economics of installing a high efficiency heat pump more favorable to the customers (Figure 13-3). The availability of alternative rates, financial incentives, and other services must be made known to the buyer during the alternative search and evaluation phase.

Figure 13-3. ROLE OF DSM IN POSITIONING HEAT PUMPS

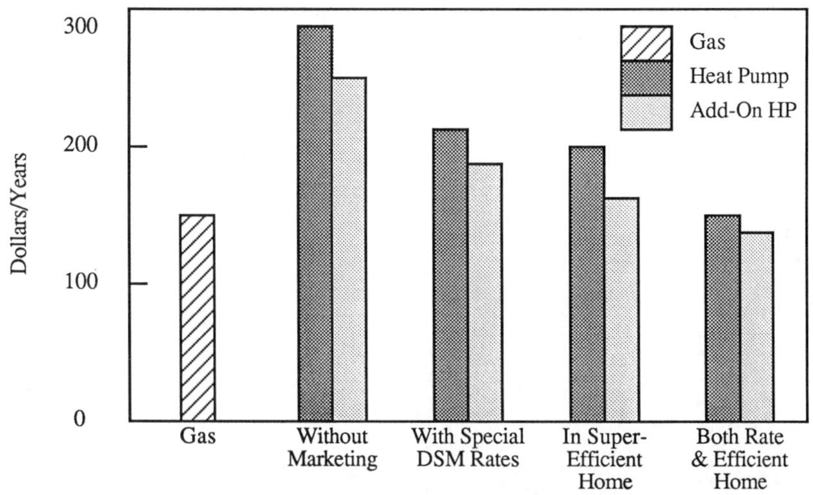

Source: Davis[5]

Many utilities have used rebates to promote heat pumps. The most successful rebate programs tend to be those in which equipment eligibility standards are designed with trade ally input. In establishing eligibility for rebates, utilities have found that they should be available to any purchaser -- whether he is a dwelling owner or a builder/contractor. A simple tiered rebate structure can be effective, with higher incentives offered for more efficient units.

Heat pumps must be properly positioned during the search and evaluation phase in order to maximize the chances that targeted market segments will react favorably. Low operating cost, performance, and efficiency, with emphasis on both the heating and cooling cycles, are three important attributes of heat pumps that should be stressed. Figures 13-4 and 13-5 illustrate the importance of heat pump attributes as stated by heat pump owners in two utility service territories. In the replacement market (types C, D, and E), low comfort levels and high operating costs are the major reasons for dissatisfaction with heating systems. This suggests that a marketing program that emphasizes comfort and cost-effectiveness be implemented to address the replacement market. Other desired attributes identified in consumer research completed by several utilities indicate that electricity would be the preferred energy source for heating if energy costs were equal to other energy sources.

Figure 13-4. REASONS FOR SATISFACTION WITH HEATING SYSTEMS

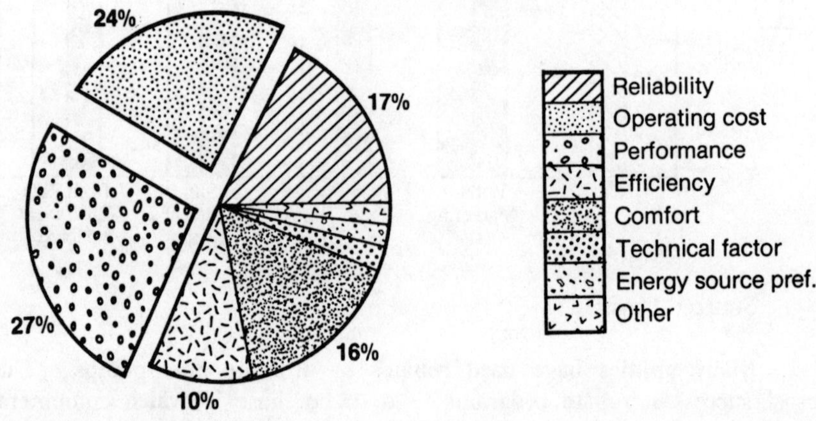

Source: Davis[5]

The replacement market is usually more difficult to penetrate than the new construction market when promoting heat pumps. In a survey completed by one electric utility, only 16% of respondents stated that they would switch energy sources to achieve the desired economics, and 40% said that they would definitely not switch energy forms. The propensity to switch to other energy forms is correlated with higher income and more dedicated customers.

Figure 13-5. REASONS FOR SATISFACTION WITH COOLING SYSTEMS

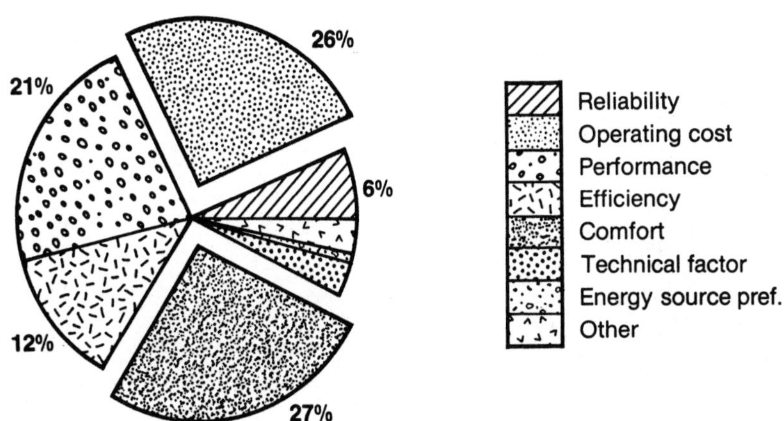

Source: Davis[5]

EPRI research on demand-side management reports customer awareness levels of heat pumps ranging from 10% to 60%. Table 13-3 rates the factors involved in residential customer decision-making for the purchase of heating systems (based on an EPRI research project[4]).

Table 13-3. RATING OF FACTORS IN RESIDENTIAL CUSTOMERS' DECISION-MAKING

	New Construction	Replacement
Reputation of dealer/manufacturer	7.7	8.8
Operating cost	6.9	7.8
Equipment cost	8.4	3.6
Fuel source	8.4	7.3
Payback (capital vs. operating)	6.6	7.1
Installation cost	7.9	6.6
Maintenance cost	6.1	6.1
Advertising	4.0	4.0
Multiple zones	4.0	3.2
Aesthetics	3.3	2.9

Legend: 1 = Least Important
10 = Most important

Actions

Once the action stage is reached, a decision has been made. For those in the type A and type B segments, a decision to install a heat pump can mean a delay in the construction of their new home(s). Each type will behave differently in following through at the action stage, and it is at this point that certain marketing programs can have an influence. Types C, D, and E sometimes need a little extra prodding to go from the alternative search and evaluation stage to the action stage; direct customer contact is often the most effective marketing technique in this situation.

When considering the action stage, the utility must first realize that there are a number of barriers which can delay or block action. Some important barriers to actions include:

- The customer cannot get cooperation and lacks the ability to override other participants who hold opposing views in the decision-making process. Obviously, in cases of multiple decision-makers, this barrier can be a substantial one. For example, a utility marketing representative has convinced the architect and builder of a new subdivision that electric heat pumps are the best choice for heating/cooling. The bank financing the builder has never loaned money on anything but gas heated homes, however, and therefore wants to earn a higher interest rate due to a perceived higher risk.

- The customer lacks adequate time or money to implement the desired action. For example, a young professional couple has decided to purchase a new condominium. Although the builder offers several options for heating/cooling systems, this couple is too busy to examine the alternatives and selects gas because that's what their parents always had in their homes.

- The customer lacks access to the means of implementation. A service the consumer wants is not offered in that community. For example, a single homeowner has decided to install a heat pump, rather than just central air, and to throw out his old furnace. There are no servicemen who have heat pump experience nearby, so he opts for a new electric air conditioner with pulse combustion furnace -- an advertised feature of a local serviceman/installer.

If the utility can exert influence to prevent barriers from occurring or weaken barriers already present, implementation of the intended action is encouraged and facilitated; the time period between decision intent and actual action can be reduced significantly.

To improve the chances that customers will purchase a heat pump when they reach the action stage, a utility should include trade allies in its marketing program. There are a number of trade allies that a utility may rely upon, including builders (for the new construction market), heating/cooling contractors, distributors, and manufacturers. These trade allies provide a vital intervening link between the utility and customer (Figure 13-6).

Figure 13-6. UTILITY/TRADE ALLY/CUSTOMER MARKETING NETWORK

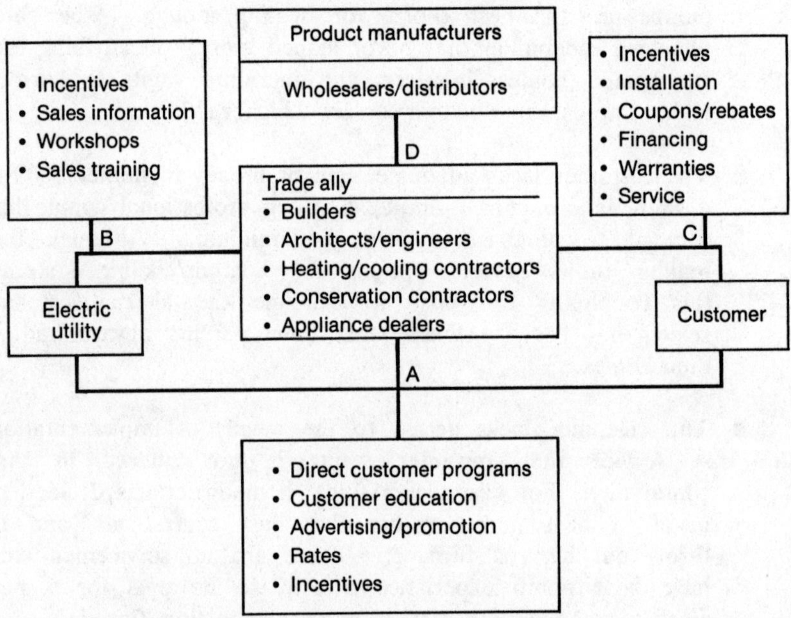

Source: Electric Power Research Institute[6]

There are a number of methods a utility can use to enlist the aid of trade allies in developing and implementing its marketing program:

- Workshops/training programs
- Lunches/dinner meetings
- Vacation trip incentive awards

- Monetary incentives
- Desk references
- Cooperative advertising
- Spiffs (jackets, caps, calendars, and other paraphernalia)
- Equipment sizing software.

Post-Action Behavior

Post-action behavior is often ignored, but as market penetration targets are met, this stage of the consumer's decision process becomes critical in terms of ensuring favorable peer recognition of heat pumps. Post-action doubts or concerns are common for consumers. Such concerns can be strong when the implementation of a decision involved high costs, financial or otherwise; when several alternatives or a close second choice alternative have been forgone; or when the alternative chosen is viewed as a mixed blessing (one that will bring significant future costs as well as benefits). The consumer wonders whether the information and feelings upon which his decision was based were adequate/accurate, whether he did the right thing.

These concerns cause the consumer to be receptive to, or to actively seek, reinforcement that the decision made was a good one. The input of others becomes influential to the consumer, and as the consequences of the action become apparent, the consumer's concerns are confirmed or dissipated. If the consumer's doubts are alleviated, the consequences of the decision will be experienced as positive and value will be increased. At this point the consumer exits the decision-making process and tends to form a habit. Such habit formation is particularly likely when pleasant experiences are immediate or strong, and clearly attributable to the decision or to the positive feedback of others in response to the decision. Faced with the same or a similar problem in the future, the consumer goes through a truncated version of the decision-making process. For instance, the consumer may skip much of the alternative search and evaluation stage and move rapidly from intent into action. The action taken will be a repeat of the previous choice: same decision, same action, but with fewer post-action doubts experienced.

In EPRI surveys, utilities have reported that participants in heat pump programs consistently report a high degree of satisfaction.

UTILITY EXPERIENCE IN MARKETING HEAT PUMPS

Heat pump marketing programs have a better chance of meeting their goals if the wealth of experience regarding similar programs that has been accumulated to date is taken into account. Many of the factors documented in utility reports and EPRI research, and others as yet undocumented, need to be considered when developing heat pump marketing programs.

Changes in the prices of competing forms of energy and improved customer access to competing energy sources have significantly influenced customer acceptance of heat pumps. As a result, many utilities have found that heat pumps are gaining greater acceptance in upscale households. However, a wide range of response levels is possible depending on the marketing strategies used. A summary of the results of a number of heat pump marketing programs is presented in Table 13-4.

Attempts to analyze utility heat pump marketing progress in order to quantify the impacts of various attributes of the program have to date been unsuccessful. There are several reasons for this lack of success. First, to properly address all of the attributes of the consumer decision process, the impact of individual program elements must be amenable to dissection. Utilities have employed a mixture of marketing techniques. Table 13-5 depicts an example of the mixture of marketing techniques employed by utilities which claim to have run successful programs. For example, analyzing the impact of radio versus television in a utility's program is, at best, a difficult task.

Table 13-4. UTILITY HEAT PUMP PROGRAMS

Utility	Program	Total Eligible Market	Total Participants	Years Programs In Effect	Average Loan/ Rebate
Detroit Edison Co.	Heat Pump Program	600 heating/cooling contractors	200	8	--
Kansas Gas & Electric Co.	Pump Up '85	70 heating/cooling contractors	60 heating/cooling contractors	1	--
		200 builders	30 builders		
Florida Power & Light Co.	Conservation Cooling	62,000 customers	36,308	4	$360
Clark County PUD	Heat Pump Financing Program	80,000 customers	43	1	$2,800
Central Illinois Light Co.	Heat Pump Rebate Program	--	--	1.5	$2,500
Buckeye Power, Inc.	Dual Fuel/Add-on Heat Pump Program	48,000 customers	200	3	$2,400/400
Portland General Electric Co.	Add-on Heat Pump Program	200,000 customers	500	6 months	$3,000
Portland General Electric Co.	A Heat Pump for All Seasons	130 customers	42	6 months	--
Salt River Project	Heat Pump Incentive Pilot Program	400,000 customers	1,154	10 months	$181
Baltimore Gas & Electric Co.	Heat Pump Conversion Program	--	--	3	--
Illinois Power Co.	Heat Pump Bonus Cash Program	--	--	1	--

Table 13-4 (continued)

Utility	Program	Total Eligible Market	Total Participants	Years Programs In Effect	Average Loan/ Rebate
Central Hudson Gas & Electric Co.	Heat Pump Dealer Programs	200 heating/cooling contractors	100	3.5	--
Niagara Mohawk Power Corp.	Residential Heat Pump Sales Program	150 heating/cooling contractors	40	4.5	--
Rochester Gas & Electric Co.	Heat Pump Promotion	30,000 customers	1,500	1	$150
Eugene Water & Electric Board	Solar & Heat Pump Water Heater Rebate Program	50,000 customers	43	1	$300
Houston Lighting & Power Co.	Heat Pump Incentive Program	46,000 customers	6,043	5	$550
Springfield Water, Light & Power Dept.	Heat Pump Program	45,000 customers	725	3	--
Gainesville Regional Utilities Board (?)	Heat Pump Rebate	22,540 customers	157	2.5	$230
Alabama Power Co.	Heat Pump Financing	20,000 customers	590	9 months	$3,100
Delmarva Power Co.	Delmarva Heat Pump Association	200 heating/cooling contractors	46	6	--
West Texas Utilities Co.	Energy Savings Plan	142,922 customers	6,508	2	$350

Table 13-5. EXAMPLE OF THE MIXTURE OF MARKETING TECHNIQUES EMPLOYED BY UTILITIES

Niagara Mohawk	City of Austin	Dallas Power & Light
Community Meetings	Community meetings	Community meetings
Newsletters	Door to door	Door to door
Workshop	Workshop	Trade ally incentives
Bill stuffers	Trade ally incentives	Television
Telephone contact	Television	Radio
Press release	Radio	Print
News stories	Print	Direct mail
Low interest loans	Direct mail	Bill stuffers
	Bill stuffers	Press releases
	Press releases	News stories
	Press conferences	Bill Credits
	News stories	Cash rebates
	Low interest loans	
	Cash rebates	

Arizona Public Service	Florida Power & Light
Workshop	Community meetings
Cooperative advertising	Newsletters
Radio	Trade ally incentives
Print	Television
Bill stuffers	Radio
Press releases	Print
Bill credits	Bill stuffers
Low interest loans	News stories
Cash rebates	Cash rebates

Source: Synergic Resources Corporation[7]

Another problem is that not all utilities have targeted their efforts at a particular market segment. Analyzing data across market segments requires information on the impact on each segment.

A third problem is that data often do not exist on income of targeted customers and on attitudes about use of discretionary funds to purchase heat pumps. Analyses are therefore often limited to comparing heat pumps without the HVAC systems. This does not give a good indication of behavior in the retrofit market.

A final factor that contributes to the difficulty of analyzing heat pump marketing programs is that utilities each gather their own heat pump data. Generally these data vary in their completeness and manner of collection across utilities; thus it is difficult to put these diverse data into one standardized format for analysis. Research into the development of transferable models of heat pump preference and purchase behavior is under way at EPRI.

REFERENCES

[1]Battelle-Columbus Division and Synergic Resources Corporation. Demand-Side Management, Volume III: Technology Alternatives and Market Implementation Methods. Report prepared for Electric Power Research Institute (EA/EM-3597), December 1984.

[2]Electric Power Research Institute. Customer Preference and Behavior: Project Overview. September 1986.

[3]Electric Power Research Institute. Strategic Planning and Marketing for Demand-Side Management: Selected Seminar Papers. EA-4308, November 1985.

[4]Market Assessment of Heat Storage Systems in Residential and Commercial Buildings. Palo Alto, California: Electric Power Research Institute. Project RP2036-18 (pending publication).

[5]Davis, Todd D. "Assessing Customer Acceptance of Heat Pump Marketing Strategies." Electric Utility Market Research Symposium, EA-4338, November 1985.

[6]Synergic Resources Corporation. Marketing Demand-Side Programs to Improve Load Factor. Palo Alto, California: Electric Power Research Institute, October 1985. EPRI Report EA-4267, Volumes 1-3.

[7]Synergic Resources Corporation. Guidebook on Residential Customer Acceptance: Designing Programs That Attract Customers. Report prepared for Electric Power Research Institute (SRC 7243-R3), December 1986.

CHAPTER 14

Marketing to the Commercial Sector

Dilip R. Limaye

INTRODUCTION

The commercial sector has always been an enigma to utility planners and marketers. Commercial customers represent a diverse mix of business activities, including office, retail, lodging, health care, food service, education, warehousing, religion, food stores, and others. Customer size ranges from large multistory office buildings and department stores to small one-story offices and retail stores, from large supermarkets to small convenience stores, and from large universities to small specialty schools (see Chapter 6, Table 6-3). These diverse business activities are characterized by different energy end uses, load shapes, and applicable technologies.

This chapter reviews some of the important characteristics of the commercial sector and discusses the approaches for developing marketing strategies tailored to this important class of customers.

IMPORTANCE OF COMMERCIAL CUSTOMER CLASS

Because of the transition in the U.S. from a manufacturing-based to a service-oriented economy, the commercial sector has been expanding and is expected to continue growing at a higher rate than either the residential or industrial sectors. Additionally, this sector contributes significantly to electric utility peak loads and offers the potential for application of many different types of demand-side programs.

Figure 14-1 illustrates the growth in commercial sector electricity consumption from 1960 to 1984 in the U.S. as a whole, and broken down into the ten Department of Energy (DOE) regions from 1970 to 1980. Figure 14-2 illustrates the shares of energy consumption. While total energy consumption has not increased appreciably in the last decade, the share of that total held by electricity has increased significantly.

Figure 14-1

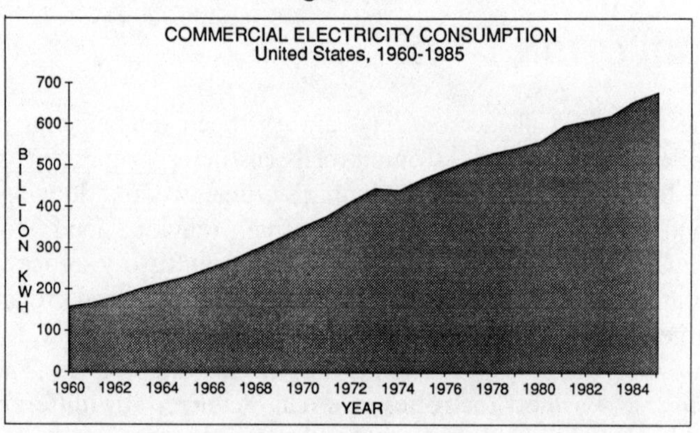

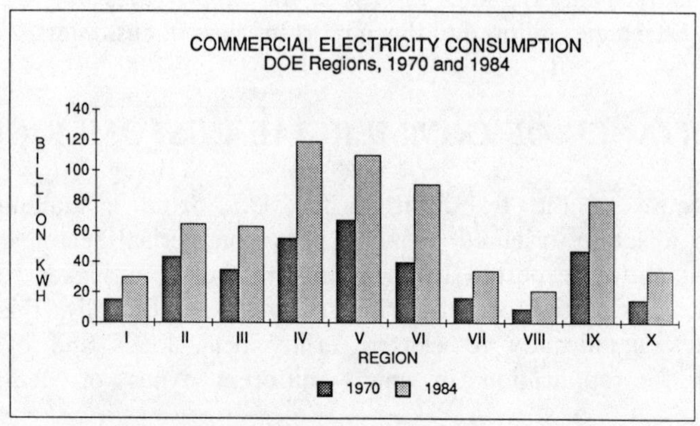

Source: Synergic Resources Corporation[1]

Figure 14-2

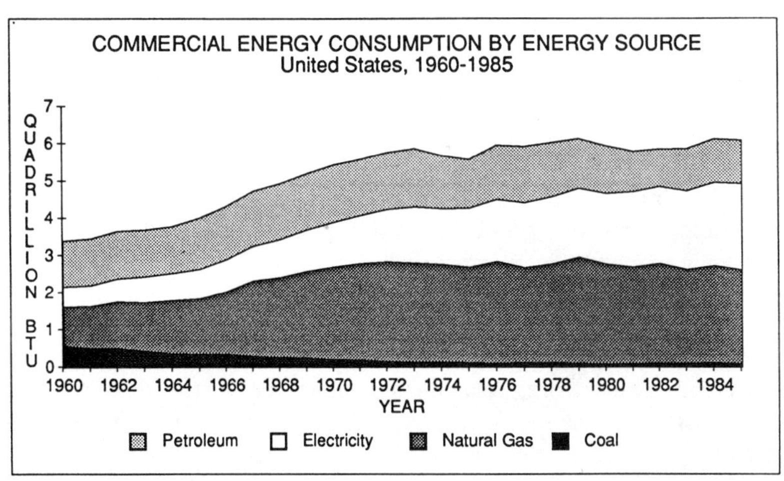

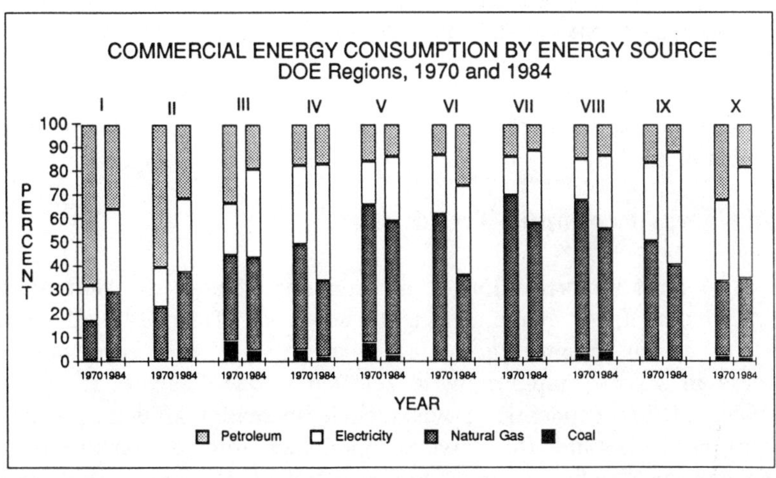

Source: Synergic Resources Corporation[1]

NEED FOR MARKET ANALYSIS

The willingness of commercial customers to participate in utility programs varies according to a number of characteristics. It is therefore essential to the marketing process to understand these characteristics and customer needs in order to communicate how a program or product meets these identified needs. Given the uncertain market conditions and the dynamic nature of the commercial sector, the development of adequate customer profile data for use in designing market strategy is imperative.

In order to develop appropriate marketing strategies and programs for the commercial sector, utilities will need to continue and expand their efforts related to commercial sector data collection and analysis. Doing so will allow utility planners to improve their understanding of the load characteristics of this sector and the major end uses and load shapes. Market analysis will also be beneficial in assessing load shaping options and identifying competitive issues, including factors which may serve as obstacles to effective marketing. Furthermore, thorough examination of this sector will benefit the utility by providing effective and efficient marketing strategy and program suggestions. The analysis of the commercial sector must include an examination of the environment in which this market is operating and some of the broad economic, socio-demographic, technological, and other trends that are likely to influence the commercial market.

Long-Term Economic Trends

The most important long-term economic trend is the shift of the U.S. economy from manufacturing to services. Table 14-1 illustrates the decline in manufacturing employment and the corresponding increase in service employment in selected business sectors from 1975 to 1984. Other economic trends include increased affluence, leading to increased demand for services; increased foreign competition in manufacturing, leading to a continuing decline of our manufacturing base; and significant increases in capital investment in commercial facilities, further assisting the growth of this sector.

Table 14-1. EMPLOYMENT TRENDS IN MANUFACTURING AND SERVICES

Business Activity	Employment (in Thousands)		% Change
Manufacturing	**1975**	**1984**	**1975-1984**
Blast furnace and basic steel	548	334	- 39%
Iron and steel foundries	230	149	- 35%
Construction and related machinery	342	257	- 24%
Industrial machinery	290	273	- 6%
Services			
Retail trade	12,645	16,584	+ 31%
Finance, insurance, and real estate	4,165	5,160	+ 24%
Business services	2,042	3,092	+ 51%
Health services	4,134	5,278	+ 28%

Source: U.S. Department of Commerce[2]

Socio-Demographic Trends

Socio-demographic trends that are likely to continue influencing the commercial sector include an increase in the number of women in the labor force (Figure 14-3), an increase in the number of two-worker families, migration to the Sunbelt states, and urban revitalization. These changes have led to increased total energy use in commercial establishments, greater contribution of commercial loads to utility summer peak loads, and an increase in commercial cooling loads, all of which are important from the electric utility perspective.

Figure 14-3. INCREASE IN NUMBER OF WORKING WOMEN (Female Labor Force as a Percentage of Female Population)

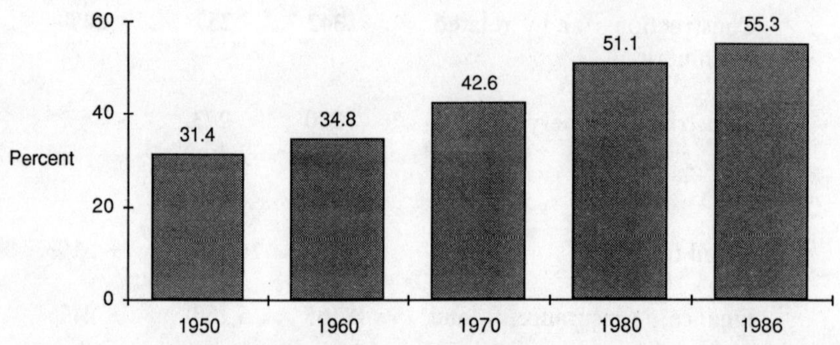

Source: U.S. Bureau of Labor Statistics[3]

Technological Trends

The development and introduction of a number of new technologies have influenced commercial sector energy consumption. Some of the important recent technological trends include:

- Office automation

- Building design and construction
- Innovative HVAC systems
- Efficient lighting
- Thermal energy storage
- Cogeneration
- Gas air conditioning.

Table 14-2 summarizes some of the implications of these technologies.

Changes in Building Ownership and Management

The changing composition of commercial ownership and management indicates that significant changes in decision-making are also underway. The salient changes in ownership and management include owners becoming increasingly remote from the buildings, increasing use of professional property managers, and more speculative building construction.

Implications

All of these changes imply that future marketing to the commercial sector may be very different than traditional marketing methods used by utilities. The commercial sector remains a very dynamic market but is perhaps the least understood customer sector. Some of the trends identified above will influence electricity demand and consumption, and utilities will need to develop adequate information on the structure, characteristics, and decision criteria to design effective marketing strategies.

Table 14-2. IMPLICATIONS OF TECHNOLOGICAL TRENDS

Technology	Implications
Office automation	Increase in miscellaneous electric loads
Building design and construction	More energy-efficient buildings; possible increase in cooling loads for buildings dominated by internal loads
HVAC systems	More energy-efficient buildings
Efficient lighting	Decrease in lighting loads
Thermal energy storage	Shifts in load from peak to off-peak periods
Cogeneration	Reduction in purchased electricity; increased gas consumption
Gas air conditioning	Reduction in purchased electricity; increased gas consumption

SEGMENTING COMMERCIAL MARKETS

Marketing to a customer sector as dynamic and diverse as the commercial class is indeed very challenging. However, this process can be facilitated by analyzing the commercial sector and disaggregating the commercial market in order to develop marketing programs. Segmentation allows a utility to tailor its products and/or services according to distinct differences in the market. Market segmentation requires a thorough understanding of the characteristics of the commercial sector. Therefore, information on the specific physical, operating, and economic characteristics, energy use patterns, and decision processes and criteria is needed.

Segmentation is generally thought of in utility marketing circles as the disaggregation of a diverse market into homogeneous subgroups for which specific utility programs can then be effectively designed and implemented. The benefits provided by market segmentation are numerous. Primarily, market segmentation enables utilities to better understand and describe customer similarities and differences, and to analyze reasons for customer decisions and behavior. Additionally, segmentation allows utilities to predict customer acceptance and response to utility programs, thereby enabling them to identify opportunities to influence customer actions. The different approaches to commercial market segmentation and illustrative examples of utility segmentation efforts are provided in Chapter 6.

END-USE AND LOAD SHAPE INFORMATION DEVELOPMENT

Developing end-use and load shape information is an important part of understanding the characteristics of a customer sector. Currently there is a lack of accurate and comprehensive data on commercial sector end uses and load shapes. As the commercial sector provides increased opportunities for electric utilities, utility planners will need information to develop strategic marketing programs that address the specific needs of this market.

End-use information is helpful in identifying markets for specific technology applications. Major commercial end uses include space heating, space cooling, ventilation, lighting, refrigeration, water heating, cooking, and miscellaneous equipment. Different end uses are important in different building types (Figure 14-4).

An ideal commercial end-use data base would contain the following information for each type of customer:

- Number of customers
- Activity or size indicator
- Square footage
- Customer listing
- Electricity consumption (kWh)
- Electricity demand (kWd)
- Total energy consumption
- Major end uses
- Customer characteristics
- Decision criteria.

Collecting such information poses several difficulties. These difficulties arise from the diversity of the commercial class, the large number of customers, the vast difference in building types and customers, multiple-use buildings, difficulties in identifying homogeneous customer sub-classes, and limited knowledge of customer end uses.

Figure 14-4. COMMERCIAL BUILDING TYPES AND END USES

Major Building Types	End Uses							
	Space Heating	Space Cooling	Vent.	Refrig.	Lighting	Water Heating	Cooking	Misc.
Office	●	●	●	—	●	○	—	●
Retail	●	●	●	—	●	—	—	○
Grocery	—	○	●	○	●	—	○	—
Restaurant	●	●	●	●	●	●	●	—
Lodging	●	●	●	●	●	●	○	○
Health Care	●	●	●	●	●	●	○	●
Education	●	●	●	—	●	○	—	—
Warehouse	○	●*	○	●*	○	—	—	—
Church	●	●	●	—	○	—	—	—

● Very Important ○ Somewhat Important — Not Important

* In Refrigerated Warehouses

Methods of collecting commercial data and developing customer load shapes vary in effectiveness. These methods include any one or a combination of the following: customer surveys, audits, secondary data sources, load research, and simulation modeling. Methods for developing load shapes include data transfer, load research end-use metering, statistical methods, econometric approaches, and engineering analysis. These methods are generally considered some of the more effective means of collecting data. An approach illustrating how the information can be collected and assimilated into an end-use data base, based on recent research sponsored by Northeast Utilities, is provided below:

- Conduct customer survey
- Obtain information on
  - Physical characteristics
  - Operating profiles
  - Energy-using equipment
  - Total energy consumption
- Classify buildings into relatively homogeneous groups
- Develop building prototypes
- Perform engineering simulation to calculate end uses and load profiles
- Validate simulation results
- Document end uses and load profiles.

An overview of this approach is shown in Figure 14-5. Typical results are illustrated in Figure 14-6.

Figure 14-5. OVERVIEW OF APPROACH FOR DEVELOPING END-USE AND LOAD PROFILE ESTIMATES

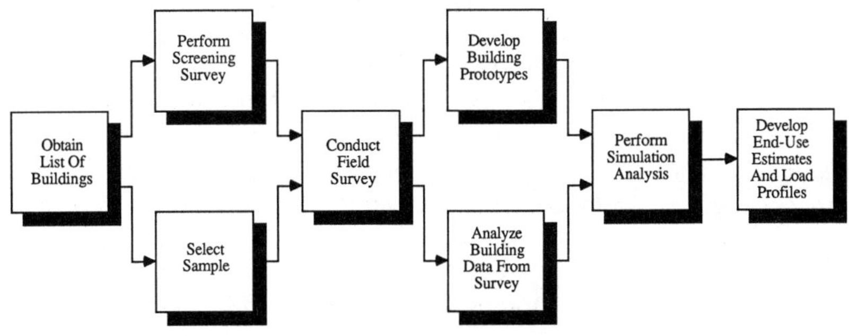

Figure 14-6. ILLUSTRATIVE EXAMPLE OF END-USE ESTIMATES

NEW OFFICE BUILDINGS

End Use		Electricity kWh/sq. ft.	Total Energy Thousand Btu /sq. ft.
Heating	-	0.1	25.6
Cooling	-	3.5	12.1
Ventilation	-	1.0	3.4
Lighting	-	6.4	21.7
DHW	-	0.3	1.2
Refrigeration	-	0.2	0.7
Cooking	-	0.1	0.5
Misc.	-	5.1	17.5
Total	-	16.7	82.7

Source: Synergic Resources Corporation[4]

TECHNOLOGIES RELEVANT TO THE COMMERCIAL SECTOR

There are numerous benefits that can be realized by both the utility and the customers from end-use technologies that offer increased energy efficiency and comfort. Examples of potential commercial technologies include office automation, energy-efficient building design and construction, high-efficiency lighting, thermal energy storage, cogeneration, and advanced HVAC systems.

The wave of the future in the commercial sector is increased office automation. The foremost example of this transition is the computer. This change will contribute to increased electricity demand in this sector. Moreover, preliminary results of end-use metering of commercial buildings indicate equipment loads have been underestimated in the past.

Contemporary building design and construction practices illustrate a growing concern with energy efficiency. Newer commercial buildings exhibit energy-efficient design, including greater use of daylighting, which reduces lighting requirements. Attention to building aesthetics is also very common -- but aesthetic design often increases the HVAC energy requirements. For example, the inclusion of atriums in commercial buildings has contributed to an increase in cooling -- and in many cases heating -- requirements in the buildings.

Increased lighting efficiency and utilization of more efficient ballasts are additional measures that have been taken to increase the overall efficiency of commercial buildings. Also, photocells are being used more frequently to control energy requirements; in terms of lighting, photocells can be used to reduce lighting loads when sunlight is available. Reduction in lighting loads also contributes to reduction in cooling loads.

Thermal energy storage is an excellent technology for utilities to market to the commercial sector. Heat and cool storage applications offer significant advantages to both utilities and commercial customers since they allow for the deferral of peak period heating and cooling loads to off-peak periods. This technology provides for peak reduction and valley filling, thereby improving load factors.

Cogeneration is another technology that may offer some benefits to electric utilities. While commercial sector growth increases a utility's sales, added peak demand and capacity requirements may not be a welcome by-product. Cogeneration is becoming increasingly attractive to commercial customers as a result of the introduction of packaged cogeneration systems. The installation of a cogeneration system is likely to have negative impacts on utility revenues and rates in the short term. In the long term, however, cogeneration may offer significant benefits through deferral of new capacity and reduced capital needs.

In addition to some of the technologies mentioned, there are many HVAC technologies that provide advantages to both electric utilities and commercial customers. Some examples include heat pumps, economizers, and absorption air conditioning. Depending on the specific utility characteristics, these technologies may be promoted to encourage either conservation or load growth.

COMMERCIAL SECTOR DECISION-MAKING

In addition to analysis of the commercial sector environment and the appropriateness of selected technologies, utility planners will need to identify the key decision-makers in commercial buildings. Once again, this information will facilitate the marketing planning process as well as increase the effectiveness of the marketing effort.

In commercial buildings, energy decisions may be made by owner-occupants, owner-developers, builder-developers, government decision-makers, architect-engineers, contractors, or tenants. It is important to know who the key decision-makers are since the factors that influence energy investment decisions will vary accordingly. Understanding the decision-makers and their respective investment criteria is essential to successful development and implementation of marketing programs.

For example, Table 14-3 shows the most important factors in the selection of HVAC systems from the perspectives of architects/engineers (A/E), developers, and builders. Table 14-4 shows the most common economic analysis methods used by owners and developers.

Table 14-3. MOST IMPORTANT FACTORS IN HVAC SYSTEM SELECTION

A/E	Owner	Developer
Reliability	Reliability	Capital cost
Efficiency	Efficiency	Reliability
	Capital cost	Efficiency
	O&M cost	O&M cost
	Proven performance	Comfort level

Table 14-4. MOST COMMON ECONOMIC ANALYSIS METHODS

Owner	Developer
Simple payback	First cost
First cost	Simple payback
Judgment/intuition	Judgment/intuition

The decisions made regarding energy investments in commercial buildings can be characterized by their complexity and their diversity. Factors which influence the decisions include the building ownership status, the building purpose, economic, aesthetic, and other criteria, and the marketing strategies employed by the utility. Some of the primary decision criteria include the following:

- First cost
- Efficiency
- Operating cost
- Useful life
- Reliability/dependability
- Adaptability
- Responsiveness
- Perceived risk.

The primary decision-makers and the process by which decisions are made are likely to vary with the size of the building. In smaller buildings, the decision-makers are generally more accessible and the decision-making process tends to be centralized and simple. In large buildings, however, the decision process can be quite complex and may involve multiple decision-makers. Many owners and developers, for example, rely on architect/engineers for advice regarding selection of HVAC systems and fuel types. The decision process in large buildings also involves numerous steps and priorities, and many factors such as aesthetics and amenities have much higher priority than energy equipment. A typical decision process for an office building is shown in Figure 14-7.

Figure 14-7. BUILDING DESIGN DECISION-MAKING PROCESS

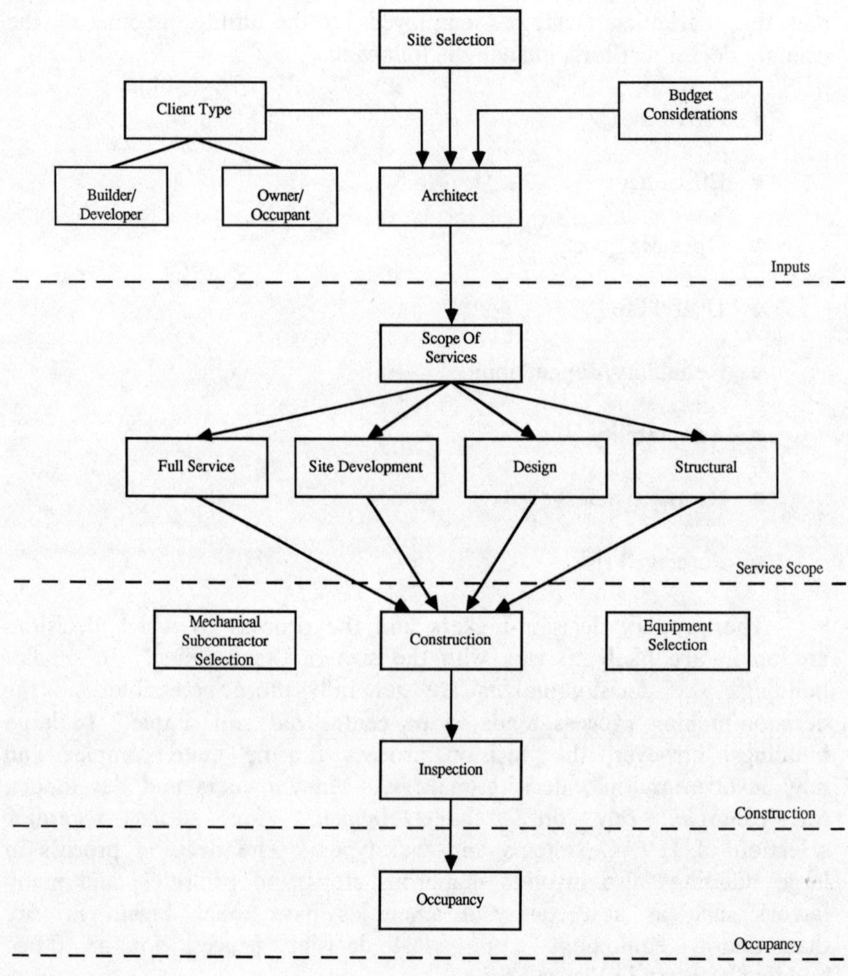

DESIGNING MARKETING PROGRAMS

The time and effort invested in the research and analysis discussed thus far will greatly facilitate the process of designing efficient and effective marketing programs. One of the most important aspects of the commercial sector is its dynamic nature, which suggests that traditional marketing approaches may be ineffective. The success of marketing strategies in today's competitive energy marketplace is directly related to the extent to which the strategy has been designed to meet current market needs. The process by which the design of optimal strategies can be accomplished includes the following tasks:

- Situation analysis
- Developing end-use and load shape data
- Segmentation
- Assessing market potential
- Evaluating utility marketing strategies
- Working with trade allies
- Developing specific plans and programs.

A number of statistical and mathematical analyses can be undertaken using commercial building market data. Advanced analysis and modeling will enable utility planners to engage in more sophisticated segmentation, to develop prototypical buildings, and to simulate building loads. Modeling allows for examination of the applicability of various technologies and investigation of the financial impacts of these technologies on different building loads. As a result of simulation analysis, attractive market segments can be determined and targeted for market penetration.

Marketing and promotional activities can then be developed specifically for individual segments of the commercial building market. Utility programs will be more successful in meeting both the objectives of the utility and the needs of the customers if the effects and technology impacts of the program have been analyzed prior to implementation. Examples of utility program/incentive effects and marketing and promotional activities for various new commercial building segments are illustrated in Tables 14-5 and 14-6.

Table 14-5. UTILITY PROGRAMS/INCENTIVES

Program/Incentive	Effect
Education/Information	Accelerated diffusion
Rebates	Reduce first cost
Low-interest loans	Reduce debt repayment
Third-party financing	Eliminate capital requirement
Performance guaranteees	Change decision criteria (?)
Rate structures	Reduce operating costs
Design assistance	Accelerate diffusion

Table 14-6. MAJOR DECISION-MAKERS IN NEW COMMERCIAL BUILDING MARKET AND APPLICABLE MARKETING AND PROMOTIONAL ACTIVITIES

Decision-Makers	Acceptance Criteria	Marketing Strategy
Owner-occupant	- Life-cycle cost - Rely on A/Es - Aesthetics - Least price sensitive	- Life-cycle cost analysis - Benefits of technologies - Information - A/E liaison
Investor-owner	- Most price sensitive - Risk averse - Functional concerns	- Rates - Financial incentives - Financial analysis - Information
Architect/engineer	- Risk averse - Responsive to client concerns - Sensitive to reliability and efficiency	- Technical seminars/ workshops - Desk references - Design subsidies - Performance demonstrations
Mechanical contractor	- Promote what they have in inventory - Standardization - Reliability	- Direct contact - Testimonials - Technical training
Lenders-financial institutions	- Risk averse - Credit worthiness - Reliability	- Direct contact - Testimonials

REFERENCES

[1]Synergic Resources Corporation. Energy End-Use Consumption Patterns and Indicators, EPRI EM-5126. Electric Power Research Institute, Palo Alto, CA, 1987.

[2]U.S. Department of Commerce. Statistical Abstract of the United States, 1986.

[3]U.S. Bureau of Labor Statistics.

[4]Synergic Resources Corporation. End-Use Energy Consumption Estimates for New Office Buildings, Final Report. Prepared for Northeast Utilities, 1985.

CHAPTER 15

Competitive Marketing for Electric Utilities: An Application of the Classic Principles of Warfare

Ahmad Faruqi

"The intensified competition that has taken place worldwide in recent years has sparked management's interest in models of military warfare."

-- Philip Kotler[1]

"War belongs. . .to the province of social life. . .[It can be likened] to business competition, which is also a conflict of human interests and activities."

-- Carl von Clausewitz[2]

INTRODUCTION

Increased competition is rapidly changing the natural monopoly status of U. S. electric utilities. As a result, utilities need to "win" in newly emerging markets, as well as defend their position in well-established markets. This chapter presents an analogy between the use of marketing as a strategic weapon and the principles of warfare.

On the surface, marketing and warfare appear to be quite dissimilar. However, on closer examination, significant similarities emerge. In both cases, "the will is directed at an animate object that <u>reacts</u>" and both consist of a "continuous interaction of opposites" and a "constant state of reciprocal action."[2] William Peacock's recent book, <u>Corporate Combat</u>, documents the parallels between events in the business and military worlds.[3] Additionally, Al Ries and Jack Trout, in Marketing Warfare,[4] and Barie James, in Business Wargames,[5] present several corroborating examples.

While electric utilities are showing renewed interest in marketing, only in a handful of cases has the concept been accepted -- much less implemented -- as a strategic doctrine. The U. S. experience in Vietnam, as analyzed by Colonel Harry Summers in On Strategy,[6] provides a powerful perspective on the importance of selecting correct strategies. Some readers may find this example offensive because it seems to compare business decisions with a war that resulted in tragic levels of death and destruction. No disrespect for those sacrifices is intended; in fact, quite the contrary. That painful military experience provides valuable lessons even in nonmilitary pursuits.

In the first chapter of On Strategy, Summers reports the following conversation, which occurred in Hanoi, April 1975:

> "You know you never defeated us on the battlefield," said the American Colonel.
>
> The North Vietnamese Colonel pondered this remark a moment. "That may be so," he replied, "but it is also irrelevant."[6]

Winning on the battlefield was irrelevant because the American forces had no grand strategy. In the absence of a governing strategic objective, tactical victories are meaningless. Similarly, if utilities can adopt a grand strategy of becoming more competitive, enlist marketing as a competitive strategy, and deploy end-use programs as tactics, they can win the battles and the war.

COMPETITIVE FORCES

The character of increased competition in the electric power business can be understood by following Porter's conceptualization.[7] Figure 15-1 presents five major types of competitive forces: buyers, other electric utilities, potential market entrants providing the same product or service, substitute products provided by firms outside the industry, and materials suppliers.

Figure 15-1. FIVE MAJOR TYPES OF COMPETITIVE FORCES

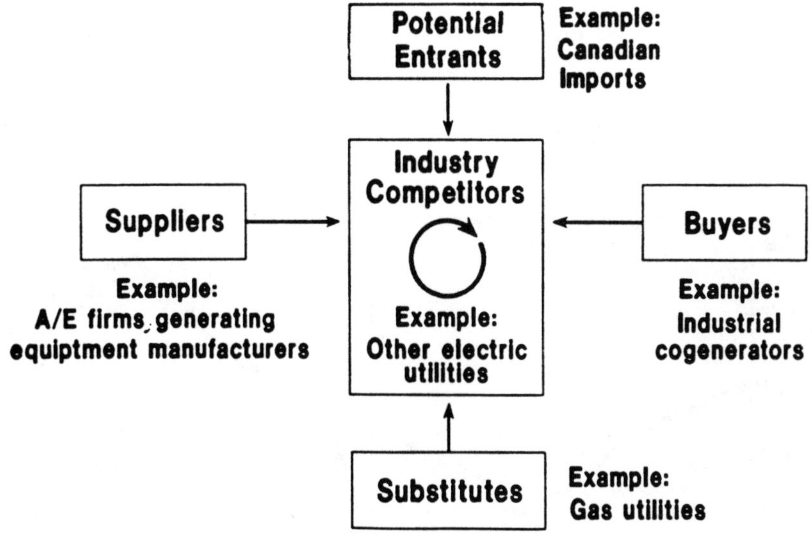

Sources: Adapted from Michael E. Porter[7] and Booz-Allen & Hamilton[8]

The first force, buyers (or customers), has not traditionally been thought of as a competitive force; however, the impact changing patterns of customer energy use have had on utility financial performance suggests that it would be wise to change the traditional way of thinking. One way of getting a feel for the effect of buyers on the industry is to look at the electric intensity of the U. S. economy; this measurement of electricity usage rose at an annual rate of 3.4% between 1960 and 1970, but slowed down in the 1970s and subsequently grew at an annual rate of only 0.6%. The main impetus for this slowdown came from the sharp trend reversal in electricity prices (from a downward trend to an upward one) that occurred in the mid-70s (Figure 15-2). Most assessments of the future outlook for electricity prices call for continued increases in real terms, making it likely that buyers -- through their reaction to such increases -- will persist as a competitive force during the next decade.

Figure 15-2. RELATIONSHIP BETWEEN ELECTRIC PRICE AND ELECTRIC INTENSITY

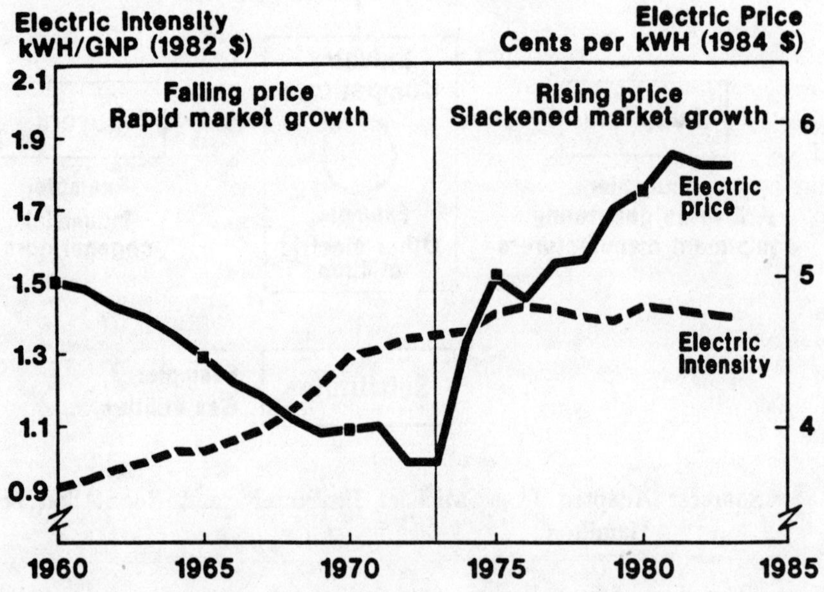

Source: Adapted from Booz-Allen & Hamilton[8]

The second competitive force is the increased role of bulk power markets. From 1970 to 1982, total utility sales grew at 3.9% annually, while wholesale sales grew at 4.8%. In 1982, wholesale sales amounted to 20% of total power sales. Moreover, interutility competition is intensifying as efforts to attract large industrial customers to new service areas become more widespread.

The third competitive force involves foreign competition, which takes two forms. The first involves direct competition in the form of electric energy imports from our neighbors; specifically, power imports from Canada and Mexico. These imports have increased their share of U. S. electricity consumption from 0.1% in 1970 to 2.0% in 1983. The second form of foreign competition is the much publicized offshore migration of electric energy-intensive U. S. industries. These industries relocate largely because of the availability of cheap electricity and labor.

The fourth competitive force involves traditional suppliers to electric utilities (manufacturers of generation equipment and consulting engineering firms). Some of these firms are now bypassing their traditional electric utility customers and hoping to become direct suppliers of electricity. Others are actively developing new technologies, such as fuel cells and cogeneration facilities, which could replace utility-supplied power.

The fifth competitive force emanates from suppliers of alternative forms of energy, primarily natural gas. That industry wants to capitalize on its price advantage and supply availability. It is also encouraging manufacturers to introduce more efficient gas-fired end-use equipment. Over the last ten years, average efficiency for gas-based residential space heating systems has grown from 67% to 90%.

Thus, the threat to electric utilities is real and perceptible on all five fronts. A timely, strategic response is called for that recognizes that utilities, like most businesses, need to focus on a triangular set of relationships involving the corporation, its competitors, and its customers (Figure 15-3).

In contemplating its position relative to these forces, electric utility management can derive significant insights from a review of the classic principles of warfare which have evolved over several centuries of actual experience. These principles find their most comprehensive expression in von Clausewitz's treatise, <u>On War</u>. This book, published in the aftermath of the Napoleonic wars, is considered "the most important philosophical study of war and warfare."[2]

Figure 15-3. THE STRATEGIC TRIANGLE

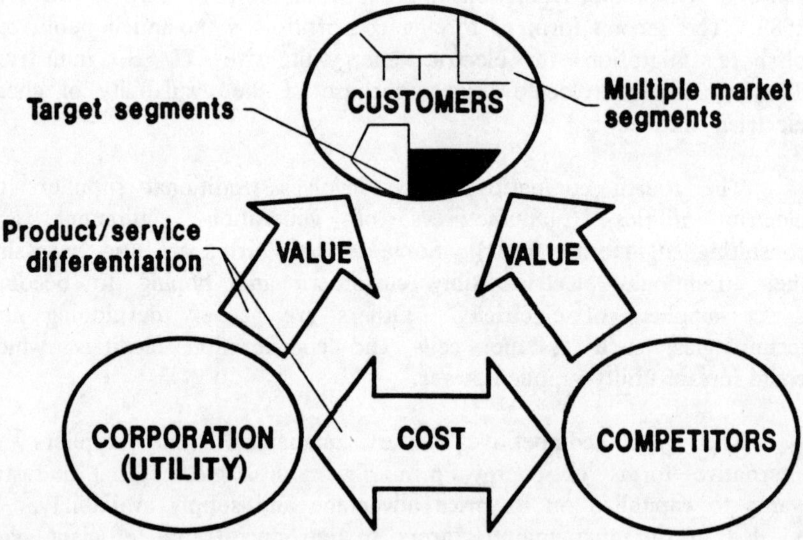

Source: Adapted from Kenichi Ohmae[9]

According to von Clausewitz, the nine principles are:

- Objective
- Offensive
- Mass
- Economy of force
- Maneuver
- Unity of command

- Security
- Surprise
- Simplicity.

For expositional purposes, this chapter posits that the utility's competitors are "the enemy," while the customer's mind (and budget) are the "battleground." The discussion format consists of four steps:

- Citation of principle
- Military example
- General business examples
- Implications for electric utilities.

Table 15-1 provides an overview.

Principle No. 1: Objective. Every Military Operation Should Be Directed Toward a Clearly Defined, Decisive, and Attainable Objective

While this statement sounds obvious, failure to observe it has all too often guaranteed defeat. Summers cites a survey conducted in 1974 which found that "almost 70% of the Army Generals who managed the [Vietnam] war were uncertain of its objective. [This mirrored] a deep-seated strategic failure: the inability of policymakers to frame tangible, obtainable goals."[6] Conversely, Israel's objectives in the Six-Day War of 1967 exemplify successful objective setting.

In general business, IBM's plans to go after Apple in the personal computer market can be cited. Clearly, IBM's reputation and size allowed entry into a market initially dominated by Apple.

Table 15-1. GRAND STRATEGY, STRATEGY, AND TACTICS

Term	Definition	Military Examples	Business Examples	Utility Examples
Grand Strategy (Statesmanship)	Policy governing the employment of war along with other national weapons	Whether or not to have a war? How to combine military resources with economic, political, and psychological weapons?	What is the corporate mission? How to position the corporation, establish key objectives?	How to respond to a more competitive environment?
Strategy (Generalship)	Use of combat for the objectives of the war	How to win the war? Which battles to fight? How to fix the place, the time, and the numerical force with which to fight the battle?	Which markets should the firm compete in? Which products or services should be developed?	How to use marketing to deal with emerging competitive threats?
Tactics (Colonelship)	Use of military forces in combat	How to win battles? How to deploy armor, artillery, and infantry units?	How to market specific products?	Which end-use programs to implement? When? For how many customers?

Source: Adapted from von Clausewitz,[2] James,[5] and Summers[6]

For electric utilities beset with significant competition, a useful strategic objective is: "Become customer- and competitor-oriented." In The Marketing Imagination, author Theodore Levitt writes, "The purpose of a business is to create and keep a customer."[10] Similarly, in In Search of Excellence: Lessons from America's Best-Run Companies, Thomas Peters and Robert Waterman Jr. present one of their main principles of excellence in business: "Get close to the customer."[11] This does not, of course, mean that other corporate functions such as production, finance, and human resources should be neglected. But it does mean making customer satisfaction the driving force behind all corporate endeavors. Utilities have traditionally been "production-oriented," i.e., driven by a desire to increase engineering efficiency and reliability. In this respect, they are similar to the Ford Motor Company in the days of Henry Ford. The company was driven then by a desire to lower production costs, while a new competitor, General Motors, decided to go after different customer segments using multiple brands. The rest is history.

Becoming customer- and competitor-oriented also means taking a long view of the future and recognizing technological change as inevitable. Railroads, of course, provide the classic example of what happens to an industry that takes a myopic view of its customers' needs.[1] Utilities need to take a broad view of the services provided by electricity. Examples from other industries include the following: Revlon manufactures cosmetics in the factory but sells hope in the marketplace; Xerox produces copying equipment but is in the business of improving office productivity; Columbia Pictures makes movies but markets entertainment; and Carrier makes air conditioners and furnaces but markets a comfortable climate in the home.

In the same way, utilities generate electricity in power plants but in the marketplace they sell energy services to make people more comfortable in their homes and offices, to enable customers to manufacture industrial products, etc. Therefore, utilities need to deploy their corporate resources to satisfy customer needs in the best possible way. Perhaps the most concise expression of customer and competitor orientation is contained in this statement coined by

Southern Company Service's Kyle Wilcutt: "Profitable production of customer satisfaction."[12]

Principle No. 2: Offensive. Seize, Retain, and Exploit the Initiative

Although the cliche "The best defense is a strong offense" still rings true, no one ever won a war through a defensive strategy. Of course, going on the offensive does not mean conducting frontal attacks at the enemy's strong points. British strategist Liddell Hart surveyed 280 military campaigns across 25 centuries and found that in only six of these campaigns "did a decisive result follow a plan of direct strategic approach to the main army of the enemy."[13] Even though such frontal attacks have rarely been successful, scores of examples exist, the most recent being the Spring '85 offensive of the Iranian army against the well-entrenched Iraqi defenses in the Basra marshes, which resulted in massive casualties for the Iranians. Other military examples of unsuccessful frontal assaults include the Battle of Bunker Hill in 1775 and the Charge of the Light Brigade at Balaklava in 1854.

Similarly, in the general business arena, RCA and GE conducted an unsuccessful frontal attack against IBM in computers. If utilities try to dissuade customers from investing in energy-efficient equipment, even when such actions are in the customers' best interests, utilities will be conducting the equivalent of a frontal attack with all the ingredients necessary for defeat. Fortunately, other, more creative strategies are available. These include: attack yourself, penetration attack, flank attack, and guerilla attack (Figure 15-4 and Table 15-2). Each of these strategies is appropriate for a specific situation: The attack yourself strategy is recommended for "market leaders," the penetration attack for "runners-up," the flank attack for "also rans," and the guerilla attack for "small entities."[4]

Figure 15-4. MILITARY STRATEGIES WHICH CAN BE ADAPTED TO THE BUSINESS ENVIRONMENT

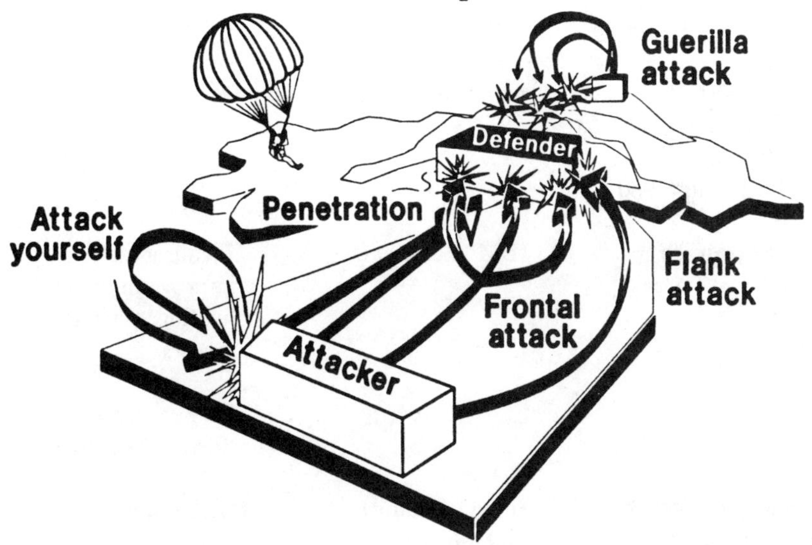

Source: Adapted from Kotler and Singh[1] and Ries and Trout[4]

Table 15-2. HOW THE MILITARY STRATEGIES CAN BE ADAPTED BY UTILITIES

THE OFFENSE PRINCIPLE

Military Example	Business Example	Utility Implications
Frontal attack ● Bunker Hill (1775) ● Balaklava (1854) ● Iran-Iraq (1980+)	Agree shampoo (1977+)	Do not fight trend toward increased energy efficiency
Attack yourself	IBM in computers	Encourage efficiency improvements in lighting, electric motors Do not adopt a "fortress" defense
Penetration attack ● Battle for France (1940)	Seiko vs Bulova, et al. (1969+)	Further develop heat pumps
Flank attack ● Alexander vs Porus (326 B.C.)	DEC vs IBM in minicomputers	Develop new industrial uses, promote storage
Guerilla attack ● Mao Zedong in China (1927-1949)	AMC vs Big 3	Electric transportation

The attack yourself strategy is appropriate in situations where the attacker has a decisive upper hand. In the military context, it involves a constant effort to stay in the leadership role by progressively introducing new weapons systems that render existing ones obsolete. In the business context, the example of IBM in mainframe and now personal computers comes to mind. Every few years, IBM launches a new series of computers that is more efficient and powerful than its predecessor. Rather than lose its business to competitors, IBM prefers to have a new product/division of IBM take business away from an old IBM product/division. This principle applies to electric utilities in all markets where electricity has a large market share -- lighting and electric motors, for example. Instead of assuming that they have an unbeatable product and adopting a "fortress" defense, utilities need to encourage more efficient methods of illumination, e.g., fluorescents, daylighting, task-specific lighting, and motion-sensors. Where motor-drive applications are involved, utilities can promote the use of variable-speed drives and more efficient motors.

The penetration attack strategies are epitomized by cases such as Seiko versus Bulova et al. and Toyota et al. versus Detroit. With the introduction of quartz-based technology by Seiko, accuracy ceased to be an issue in the marketing of watches. Accordingly, in 1969 Seiko shifted the marketing emphasis toward consumer choice by offering a lineup featuring hundreds of different shapes and sizes. Gradually, Seiko acquired a very large share of the world watch market. In the 1970s, Toyota et al. focused on the production of fuel-efficient cars and captured the low end of the U. S. automobile market, which at the time was characterized by little competition. Then, by successfully offering upscale compact and mid-sized automobiles, as well as pickup trucks, they converted their initial foothold into a steadily expanding share of the U. S. market.

Many penetration attack strategies can be employed by electric utilities. The most desirable market segments for penetration attack involve end uses for which other nonelectric alternatives are easily available and competitively priced. Prime examples are space heating, water heating, and certain process heating applications. The penetration attack strategy is strategic load growth.

Flank attacks are most appropriate if the attacker is "small" relative to the opponent. A military example illustrates the point. In 326 B.C., King Porus of India, with a vast army of 35,000 infantry and cavalry troops backed by 200 elephants, was camped on one side of the river Jhelum, which runs through what is now Pakistan. Alexander the Great made several forays on the other side of the river to "pin down" Porus' vast army. Then, early one morning, Alexander crossed the river with a small party under cover of a thunderstorm. Surprised and shocked by this attack to his rear, Porus surrendered after a few hours of battle. The flank attack strategy has its parallels in the business principle of leverage. The Japanese frequently use this principle. They find elements of a product line that a U. S. manufacturer has neglected or not yet developed. They enter with a superior, underpriced product and capture a large market share. Then they expand the product line to capture the more competitive items. Digital Equipment Corporation (DEC) used this principle when it sensed a weakness in IBM's product line (minicomputers) and launched a line of products which enjoyed remarkable success.

A flank attack strategy for utilities could involve the following steps: identifying shifts in market segments that are creating gaps in competitors' product lines; rushing in to fill the gaps; and developing them into strong segments. Consider the manufacturing sector. Virtually all manufacturers are trying to improve productivity so they can remain competitive in world markets. The actual cost of electricity in several major industries is a small fraction of production costs, with labor and materials accounting for the bulk of such costs. However, there are several electrotechnologies available that can improve industrial productivity substantially by reducing processing time, material costs, and/or labor costs; there is a tremendous potential for utilities to market such technologies.[14] Similar opportunities might be pursued in other market segments; low, off-peak power rates could be used to encourage the penetration of thermal energy storage technologies, for example.

Several examples can be cited to support the industrial productivity observation. Intel Corporation recently established a semiconductor assembly plant in the United States, breaking from the tradition of siting assembly plants offshore. The rationale provided by Chairman Moore: "We are able to build a cost-effective component assembly plant in the United States for the first time because the assembly process is becoming increasingly automated and therefore less labor-intensive."[15] President Grove added: "One Intel assembly line operator in 1985 will be able to do the work done by 180 operators in a 1977 Intel plant. Obviously, this will have a great effect on the labor cost component in everything we produce."[15] In another example, General Motors Chairman McDonald indicated recently that the "technology boom" in manufacturing could bring jobs back from overseas to the United States. He cited the example of circuit boards once made by GM in Singapore that are now being made in the United States because advances in technology outweighed the higher U. S. labor costs. "Beat the lower cost with technology. We think we can do that time and time again," McDonald said.[16]

Guerilla attack strategies are appropriate when the attacker is very small compared to the defender. A successful military example of this tactic is the eventual victory of the Chinese communists under Mao Zedong. From 1927 to 1949, Mao's guerillas forced first the Japanese and then the nationalist Chinese off the mainland. Mao's philosophy was: "The enemy advances, we retreat. The enemy camps, we harass. The enemy tires, we attack. The enemy retreats, we pursue."[2]

A business parallel to the guerilla attack is found, once again, in the automobile industry. AMC, for example, has created a niche in the Jeep segment, where the "Big 3" offer no effective competition. Similarly, electric utilities, through the electric vehicle (EV) product, have the potential to mount a guerilla attack in automobile markets. The EV has several attributes which make it attractive for fleet-type applications (delivery van, shuttle bus, mail van). With the availability of low, off-peak electric rates (needed to reduce the cost of the overnight recharging period), this vehicle could become a significant new use for electricity.

Principle No. 3: Mass. Concentrate Combat Power at the Decisive Place and Time

The first principle of war, the objective, focuses on what needs to be accomplished. The second principle, the offensive, shows how the objective can be attained. The next three principles -- mass, economy of force, and maneuver -- elaborate on the second principle by providing more details on how to conduct military operations.

Liddell Hart said that "the principles of war -- not merely one principle -- can be condensed in a single word, concentration, but for truth this needs to be amplified as the concentration of strength against weakness."[13] Von Clausewitz states that superiority of numbers is "in tactics, as well as in strategy, the most general principle of victory. Where an absolute superiority is not attainable...produce a relative one at the decisive point."[2] He attributes the successes of Frederick the Great and Napoleon Bonaparte to their resolution in keeping the "forces concentrated in an overpowering mass."[2] In all military cases, superiority of numbers at the decisive place and time ensured victory for the attacker.

As an example from the business world, IBM successfully fought off the attack by Exxon and Lanier in the office automation market largely based on the mass principle. Similarly, Swanson fought off the attack by Stouffer et al. in the TV dinner market.

Electric utilities often deal with different types of customers, each facing multiple influences. These influences have to be analyzed to reveal the points in the buying process at which the influence of utility programs is the strongest and the influence of competitors' programs the weakest. After all, the whole art of warfare is to deploy one's own resources at the point of the competitor's weakness, i.e., positioning.

The key to positioning is strategic reconnaissance, which identifies competitor strengths and weaknesses and uncovers changes in customer needs. One way to conduct reconnaissance is to segment energy markets based on customer needs and homogeneity of customer responsiveness to utility programs. This requires the development of market intelligence, i.e., demand forecasts disaggregated by end uses. Such intelligence can provide insights into customer behavior in several "what if" scenarios. It may suggest pricing electricity according to value-of-service rather than traditional cost-of-service considerations.

Analysis of the iron and steel industry reveals, for example, that the share of electricity in total energy use is expected to rise (Figure 15-5). To assess the leverage utilities have in this share, two sensitivity cases are examined, one with "free" electricity and another with electric price growth at twice the base case. The free electricity case illustrates the upper limits of the electric share based on technological constraints, and the expensive electricity case provides a lower bound. Similar simulations can be performed with respect to variations in the price of alternative fuels, economic conditions, etc. Comparisons across these cases will indicate which factors have the greatest influence on electricity use. This type of information, generated across several market segments, can identify segments in which utility programs have the greatest leverage. Once this information is combined with the costs of program implementation to produce a net assessment of competitive strength, utilities will be able to determine where to concentrate their corporate resources. Based on such logic, several utilities are currently offering "economic development" rates. Such pricing tactics boost the utilities' competitiveness by attracting highly profitable industrial loads.

Figure 15-5. SHARE OF PURCHASED ELECTRICITY IN PURCHASED ENERGY, IRON AND STEEL INDUSTRY

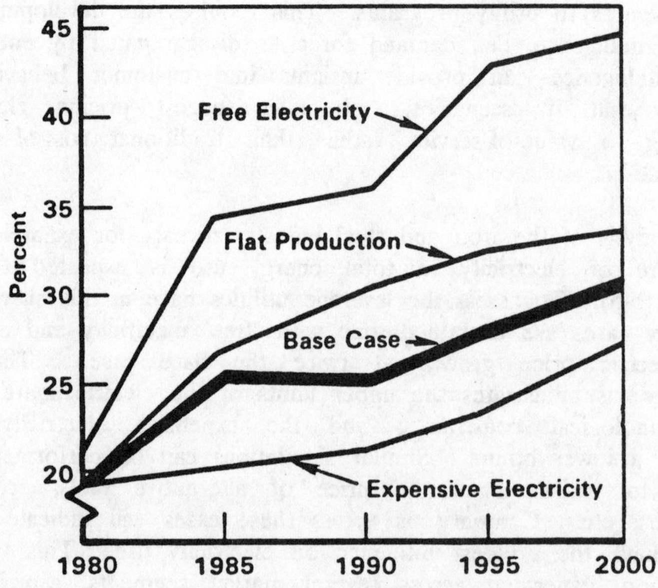

Source: Adapted from Brookhaven National Laboratory[17]

Principle No. 4: Economy of Force. Allocate Minimum Essential Combat Power to Secondary Efforts

Economy is the reciprocal of mass. Because absolute superiority everywhere is unattainable, relative superiority somewhere is recommended under the mass principle. This is often achieved at the price of relative inferiority in other areas. This principle counsels that inferiority, far from being a handicap, is a prerequisite to strategic success. Indeed, as Liddell Hart observed, "True concentration is the fruit of calculated dispersion."[13] In several instances, failure to follow this principle has guaranteed defeat. Frederick the Great said, "He who defends everywhere defends nothing."[13] Reviewing his marshals' plans for defending France,

Napoleon found that la Grande Armee had been dispersed to defend every kilometer of the empire's boundaries. He chastised the marshals for selecting an unattainable objective, saying that there were "not enough men in the whole of France" to permit its realization, and that such dispersion was the surest harbinger of defeat.

In the business arena, examples of this principle include recourse to bankruptcy proceedings, personnel raids on rival firms, take-over of efficient competitors, and legal action against die-hard rivals.

Many electric utilities are faced with a constrained marketing budget and multiple objectives. Multiple objectives may include: satisfaction of regulatory pressure to successfully promote a low-income conservation program; increasing kWh sales; and improvement of the utility's image to cut down on customer complaints. With a budget that is often constrained, a utility is forced to spread programs so thin that the overall effort is rendered ineffective. Thus, if a utility is to achieve its most important marketing objectives, it must devote an adequate budget to primary data collection and develop a mix of marketing programs that best matches company capabilities with unsatisfied customer needs. Invariably, this will mean accepting prudent risks by allocating fewer resources for secondary objectives.

Principle No. 5: Maneuver. Place the Enemy in a Position of Disadvantage Through the Flexible Application of Combat Power

The indirect approach often brings success. The Japanese conquest of the "impregnable fortress" of Singapore in 1941-1942 was achieved by cutting through supposedly impassable tropical jungles and attacking the fortress from the rear, completely obviating the need to cope with its vaunted naval battery equipped with 15-inch guns.

Maneuver can also be applied by forcing the enemy to attack where one's own defense is strongest. This has the effect of turning dangerous flanking attacks into harmless frontal attacks. This strategy is described by Helmut von Moltke in the following passages:

> A clever military leader will succeed in many cases in choosing defensive positions of such an offensive nature from the strategic point of view that the enemy is compelled to attack us in them.
>
> The task of a skillful offensive will consist of forcing our foe to attack a position chosen by us, and only when casualties, demoralization, and exhaustion have drained his strength will we ourselves take up the tactical offensive...Our strategy must be offensive, our tactics defensive.[13]

An excellent business example of this type of maneuver is provided by the competition in the market for greeting cards. American Greetings was a newcomer facing an uphill battle against the well-entrenched Hallmark. American Greetings resorted to creating the Strawberry Shortcake doll, sold it as a toy, and, when the toy was well established, used the now well-known symbol to successfully promote its cards. Another example is the rejuvenation of the consumer radio market by Japanese manufacturers. While American producers had given up on the radio market, viewing it as saturated, the Japanese stepped in with an innovative array of variations on the basic radio, as shown in Figure 15-6.

In the case of electricity, the role electrotechnologies can play in improving productivity should be considered. Visualize a situation where, in response to targeted utility marketing, an industrial plant installs more electric-intensive equipment. This alone should increase electric usage. However, because the plant will now be more productive, output will also rise, leading to a further rise in electric usage. The indirect effect of the increased output on electric usage will often be stronger than the direct effect of increased electric intensity. This suggests that the indirect approach may be advantageous even in utility marketing.

Figure 15-6. EXAMPLE OF THE REJUVENATION OF A SEEMINGLY SATURATED MARKET

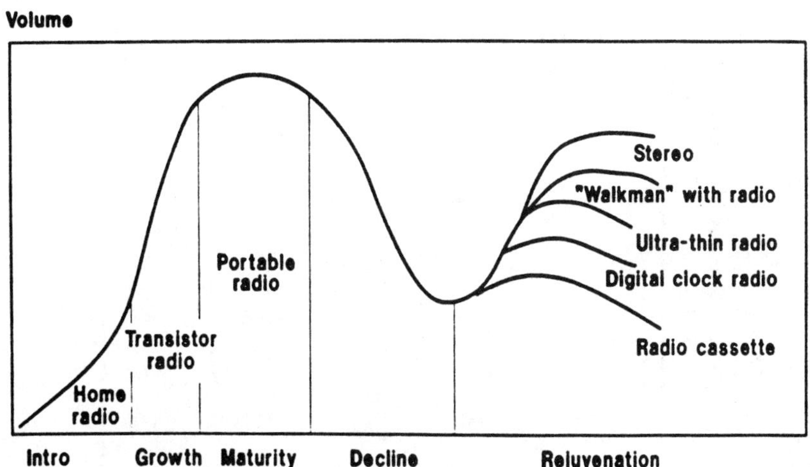

Source: Adapted from Kotler and Singh[1]

Utilities can, for example, develop new end uses to compensate for the saturation of existing ones; i.e., become less concerned with market share and more with market creation. Utilities can also rejuvenate mature products through imaginative marketing. This may involve offering customers a variety of pricing options, developed through tools such as those shown in Figure 15-7. The circles correspond to a baseline case, and the arrows show how industrial electric use may be optimized through utility programs that either increase the growth rate of electric-intensive industries or help electrify fast-growing ones.

Figure 15-7. INDUSTRIAL MARKET PLANNING MATRIX

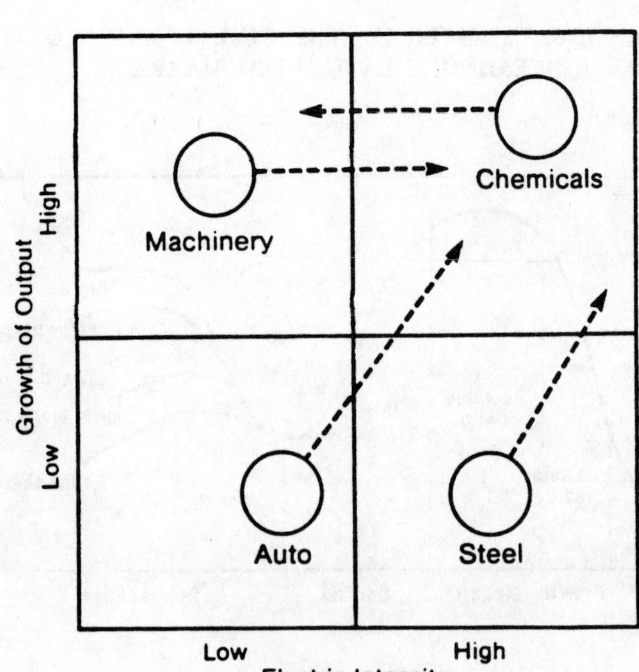

Source: Ahmad Faruqi et al.[18]

Principle No. 6: Unity of Command. For Every Objective, There Should Be Unity of Effort Under One Responsible Commander

Violation of this principle led to the Battle of the Bulge in 1944. Hitler's commanders sensed a lack of unified command among the Allies and launched a counterattack through the Ardennes. This desperate counterthrust resulted in massive Allied casualties and almost gained a reprieve for the Axis powers. The Allied command had failed to anticipate an attack in a place where just four years earlier Guderian's panzer forces had made the word blitzkrieg part of the English lexicon; the Allies had fallen victim to "matrix" management. In an effort to keep unanimity among all countries, each was given a portion of command responsibility, resulting in a weakened defense.

In general business, several corporations are abandoning "matrix" management approaches for similar reasons. Firms such as Texas Instruments, International Telephone and Telegraph, and General Electric were once ardent believers in matrix management. They are now convinced that blurred lines of responsibility lead to matrix mismanagement.

Similarly, some utilities are beginning to incorporate demand-side management alternatives in their resource planning process, thereby introducing an element of unity into their corporate management efforts. Others have developed detailed end-use assessments of market segments, using information from a variety of demand-side departments such as load research, load forecasting, and customer service. But very few utilities have developed a truly integrated planning capability. To do so may involve implementation of organizational changes, such as incorporating the rate-making function into the marketing department; removing wholesale power sales from the production department; and making utility-wide training programs consistent with a marketing orientation.

Principle No. 7: Security. Never Permit the Enemy to Acquire an Unexpected Advantage

Security prevents the competition from conducting surprise attacks and from turning flank attacks into frontal offensives. Military examples of the violation of this principle are legion, and include the Homeric legend of the Trojan Horse; the Japanese attack at Pearl Harbor in 1941; and Germany's Russian campaign in 1941-42. In business, IBM's clandestine development of the personal computer demonstrates the benefits of this principle.

This principle suggests that utilities need to monitor and analyze competitor behavior so as to discern the competitors' future course of action; utilities must develop early warning systems. Utilities also need to study evolving changes in customer preferences and behavior. To do this effectively, they need to hire key people with strong marketing backgrounds and establish strong market research departments. They also need to investigate the impact major structural changes in the industrial sector will have on utility sales, revenues, and profitability, and develop strategies for shaping the future pattern of loads. This may involve understanding not only the needs of the utility's industrial customer but also the needs of the industrial customer's client, especially for industries competing in the international environment. Ultimately, this may involve fostering the development of demand-side "champions" within the company, i.e., senior executives committed to articulating and implementing end-use strategies.

Principle No. 8: Surprise. Strike the Enemy at a Time, Place, and in a Manner for Which He is Unprepared

Surprise is the reciprocal of security. An oft-cited military maxim claims "a single act of surprise is worth a thousand soldiers." While strategic surprise is very difficult to achieve, tactical surprises are often possible. However, prompt follow-up is required to fully exploit the benefits of surprise.

Intelligence about the enemy's intentions and disposition of forces is a critical prerequisite for achieving surprise. Such intelligence should measure not only the easily quantified variables that indicate enemy strengths and weaknesses, but also the intangible, "psychological forces and effects" analyzed by von Clausewitz.[2] Summers cites a relevant yet typically bizarre story from the closing days of the Vietnam War:

> When the Nixon administration took over in 1969 all the data on North Vietnam and on the United States was fed into a Pentagon computer -- population, gross national product...number of tanks, ships...and the like.

> The computer was asked, "When will we win?" It took only a moment to give the answer: "You won in 1964!"[6]

On the topic of surprise, the Chinese strategist Sun Tzu wrote in 500 B.C.:

> All warfare is based on deception. Hence, when able to attack, we must seem unable; when using our forces, we must seem inactive; when we are near, we must make the enemy believe that we are far away; when far away, we must make him believe we are near. Hold our baits to entice the enemy. Feign disorder, and crush him.[19]

Military examples of surprise include Hannibal versus the Romans at Cannae (216 B.C.), Napoleon versus the Austrians and Russians at Austerlitz (1805), the Germans versus the Russians at Tannenburg (1914), the Germans versus the French at Sedan (1940), the Allied landing at Normandy (1944), and the Israelis versus the Arabs (1967).

In many ways, the achievement of surprise requires two essential elements: "calculated dispersion," i.e., keeping the enemy uncertain about one's true objectives by playing up several alternative objectives, thereby dividing the enemy's forces; and executing the true plan with utmost speed. As the Napoleonic creed goes, "l'audace; l'audace; toujours l'audace (speed; speed; and always speed)." It is germane to quote from Guderian on the importance of movement and speed:

> Everything is therefore dependent on this: to be able to move faster than has hitherto been done; to keep moving despite the enemy's defensive fire and thus to make it harder for him to build up fresh defensive positions; and finally to carry the attack deep into the enemy's defenses.[20]

Meade's failure to pursue the defeated Lee at Gettysburg (1863) provides an illustration of the consequences of failing to exploit the advantage of surprise; the Civil War dragged on for another two years.

In business, the successful introduction of the Beetle by Volkswagen in the United States in the 1950s is a good example of surprise. Many analysts regarded the Beetle as unlikely to succeed because of its odd shape, small size, and, most of all, its public image as a staff car in Hitler's army. However, Volkswagen offered a service contract to its buyers -- an unheard of arrangement at the time -- and proceeded to carve a niche for itself by promoting the concept of "thinking small."

For a regulated utility, complete surprise may be difficult to achieve. However, the emphasis on speed may be worth keeping in mind. An example of this idea is the electric utility's positioning with regard to the threat presented by small, packaged, gas-fired cogeneration devices. Rather than stand by and watch the gas utilities walk away with premium load, electric utilities could:

- Attempt to change rates so as to put standby and/or load factor rates in place now, making the installation of small cogeneration packages less desirable to customers
- Aggressively market high-efficiency packaged chillers and other devices that make it difficult for small cogeneration to compete
- Develop a small cogeneration business.

Principle No. 9: Simplicity. Keep All Plans Simple Because Even the Simplest of Plans is Difficult to Execute

The 1982 British campaign in the Falklands was a contemporary application of this principle. In business, the divestiture activities of large corporations such as Dupont, Beatrice Foods, and Firestone Tire and Rubber exemplify the need to simplify. In each case, the companies sold subsidiaries that were not directly related to their central business in order to consolidate resources and reduce the resistance of the marketplace.

Marketing, like warfare, involves effort directed through a resistance. Anything that can be done to reduce the "friction in the machine" should be implemented to prevent the "fog of war" from paralyzing operations. One of the methods suggested by von Clausewitz to "diminish the natural friction" and thereby improve the "readiness, precision and firmness...in the movement of troops" is to conduct "constant repetitions" of "formal exercises" by which the troops get habituated to action.

Utilities need to keep this principle in mind as they design their demand-side plans. A good method for ensuring simplicity (and also for keeping the organization's readiness level high) is conducting pilot tests and analytic simulations prior to large-scale implementations.

CONCLUSIONS

Marketing offers electric utilities an effective method for combating increased competition in the end-use markets for electric power services. Based on concentration, maneuver, and speed, utilities can mount a strategic offensive aimed at the competition. Such a campaign may involve several elements.

Utilities must become customer- and competitor-oriented throughout their organizations. They must set up a command structure to integrate demand-side and supply-side options into a unified business plan, seize the initiative from the competition by exploiting the unique attributes of electricity in a few strategically selected market segments, and conduct penetration attacks.

To effect this strategy, they need to develop new uses of electricity to compensate for the saturation and efficiency of existing uses, to conduct flank attacks and guerilla attacks, and to offer a menu of pricing options based on the differing needs of customers in separate market segments.

REFERENCES

[1]Philip Kotler and Ravi Singh. "Marketing Warfare in the 1980s." Journal of Business Strategy, Winter 1981, pp. 30-41.

[2]General Carl von Clausewitz. On War. England: Penguin Books Ltd., 1968.

[3]William E. Peacock. Corporate Combat. New York: Facts on File Publications, 1984.

[4]Al Ries and Jack Trout. Marketing Warfare. New York: McGraw-Hill, 1986.

[5]Barie James. Business Wargames. England: Penguin Books Ltd., 1985.

[6]Colonel Harry G. Summers. On Strategy. Novato, California: Presidio Press, 1982.

[7]Michael E. Porter. Competitive Strategy: Techniques for Analyzing Industries and Competitors. New York: The Free Press, 1980.

[8]Booz-Allen & Hamilton. "Emerging Competition in Electricity Markets: A Discussion Document." Electric Power Research Institute, 1985.

[9]Kenichi Ohmae. The Mind of the Strategist. New York: McGraw-Hill, 1982.

[10]Theodore Levitt. The Marketing Imagination. New York: The Free Press, 1983.

[11]Thomas J. Peters and Robert H. Waterman Jr. In Search of Excellence: Lessons from America's Best-Run Companies. New York: Harper & Row, 1982.

[12]Battelle-Columbus and Synergic Resources Corporation.

[13]Captain B. H. Liddell Hart. Strategy. New York: Frederick A. Praeger Publishers, 1967.

[14]Philip S. Schmidt. Electricity and Industrial Productivity: A Technical and Economic Perspective. Elmsford, New York: Pergamon Press, 1984.

[15]San Jose Mercury, March 29, 1984.

[16]San Jose Mercury, May 26, 1984.

[17]Brookhaven National Laboratory. Industrial Process Models of Electricity Demand. Palo Alto, California: Electric Power Research Institute, EA-3507, 1984.

[18]Ahmad Faruqi et al. "Ten Propositions in Modeling Industrial Electricity Demand." In Forecasting U. S. Electricity Demand, ed. A. M. Bolet. Westview Press, 1985.

[19]Sun Tzu. The Art of War. New York: Oxford University Press, 1963.

[20]General Heinz Guderian. Panzer Leader. Washington, D.C.: Zenger Publishing, 1952.

CHAPTER 16

Moving Beyond Marketing Myopia: The Responsibility Rests with Senior Management

Todd D. Davis

BACKGROUND

Senior managers and executives at electric utilities need to periodically review Theodore Levitt's article "Marketing Myopia."[1] This article is viewed as a classic in the marketing literature and continues to be cited frequently in current marketing publications. Levitt identifies some of the key problems facing all industries, including electric utilities, and prescribes a general method for overcoming myopia and the negative consequences of its end result: the obsolescence and death of an industry. This chapter suggests that those in senior utility management have the ultimate responsibility for overcoming marketing myopia. Without the commitment of senior management, few of the principles and practices of strategic marketing presented in this book can be successfully implemented.

THE SOURCE OF MYOPIA

According to Levitt, marketing myopia stems from the following misconceptions:

- The industry is self-deceived into anticipating endless growth.

- There is a belief that there is no competitive substitute.

- Mass production and economies of scale are believed to be important focal points because of declining unit costs.

- Preoccupation with the product is assumed to lead to endless cost reduction.

A major portion of Levitt's article is devoted to arguing against these highly cherished beliefs. In support of his arguments, attention is devoted to the petroleum, railroad, and electric utility industries as examples. A lucid point made in the article is that "there is no guarantee against product obsolescence. If a company's own research does not make it obsolete, another's will."[1]

The myopia facing an industry or firm (according to Levitt) springs from the fact that a firm is product-driven and often fails to consider additional product/market strategies to extend the product life cycle. As a result of this narrow focus, customers eventually become wiser and more knowledgeable about a product. At the same time new firms enter the market and provide additional product enhancements and lower cost options. These improvements may be applied to an existing generic product or achieved through the introduction of a totally new innovation.

EXAMPLES OF MYOPIC MISTAKES

The marketing literature is replete with mistakes made by firms that failed to realize that product innovations introduced by competitors were the wave of the future:

- National Cash Register (NCR) continued to rely on electromechanical cash registers when new suppliers were moving toward electronic products, resulting in NCR write-offs of $140 million.

- B. F. Goodrich continued to emphasize bias ply and other traditional tire cord designs while Michelin's steel-belted radial tires swept the market.

- RCA failed to capitalize on its position in the vacuum tube market and transfer into the semiconductor market. From 1955 to 1975, the vacuum tube market died and so did RCA's business.[2]

There are other examples of new firms that revolutionized industries: Apple Computers in the personal computer market, Southwest Airlines in the discount airfare market, Sony in the small consumer electronics industry.

COPING WITH CHANGE

An industry that is currently going through a rapid period of change is telecommunications. In 1955, AT&T's dominance in telecommunications was unquestioned and its future seemed unchallenged. By 1970, a number of substitute products and competitors began to appear. Pressures for destandardization began to appear in the terminal market, the transmission market, and in message origination. Alvin Toffler noted the parallelism between the internal changes in the communications industry and changes in the external environment.[3] In advising AT&T on how to cope with these changes, Toffler suggested the following steps:

- A new corporate culture is needed to reflect product innovation and external change.

- Less hierarchical bureaucracy and more temporary organizations such as task forces are needed in order to respond more quickly to change.

- There must be recognition of the limitations of economies of scale.

- More importance must be attached to responding to the psychological needs of customers.

- There should be greater emphasis on environmental tracking of trends and behavioral research.

- Employee training on task completion and corporate culture is needed.

- There is an important need for total integrated planning.

- More monitoring and tracking of product/service offerings are required.

Toffler advised AT&T that "in the past, the company that knew how to standardize most effectively was able to beat out its competitors. In the future, the company that knows how to destandardize effectively may prove the victor."[3]

PRESSURES FACING ELECTRIC UTILITIES

The electric utility industry has been misled by its technological ability and the importance of economies of scale. While the current electronics revolution continues to run its course, there are many signs that some customer groups want more service flexibility in terms of suppliers, rate schedules, and service levels. It could be argued that similar pressures for destandardizing electric service are occurring in the electric utility industry. Moreover, a number of competitive substitutes are beginning to emerge and challenge the economies of scale involving central station generation. For example, non-utility generation represents a significant form of competition for utilities because it opens up the power supply business to medium-to-large commercial/industrial electric customers. Moreover, a number of third party financing firms are emerging that are providing low cost capital to those wishing to invest in alternate forms of electric supply.

The preoccupation with the traditional notion of supplying electricity continues in most electric utilities. Many electric utilities have skewed staffing and other resource commitments in favor of the supply side of the business. Few utilities have resources devoted to creating new product/service innovations. While most utilities today report having no plans to build any additional baseload plants until the next century, indecision seems to prevail regarding the role of demand-side management, marketing, and innovation. Only a few electric utilities have created new product and new business development areas within their organizations.

VIEWING AND USING THE PRODUCT LIFE-CYCLE CURVE

A fundamental principle behind Levitt's article and one that continues to be important in today's marketing literature is the product life cycle. The product life cycle is important because it suggests how an organization should modify or adapt its marketing mix over time. Moreover, it suggests that organizations should not be caught with too many products in one phase of the life cycle. If all products are in the introductory phase, then cash flow and financial viability could be a problem. If all products are in the latter phases of the life cycle, the viability of the future of the business could be jeopardized. Usually the early phases of the life cycle are very slow, followed by rapid growth in product enhancements and/or market share, which then reach a limit in technical progress that leads toward market share declines. For instance, ships do not run faster, washers and dryers do not wash and dry clothes much cleaner and faster, and the heat rates of power plants do not get any better. These limitations are inherent in the technology and, in the end, influence market acceptance.

Using the product life cycle when monitoring sales and considering marketing strategies is essential, given the destandardization or discontinuity which threatens the very existence of a firm.[4] Whereas one competitor may be nearing its limits in terms of technological development, price, and competitiveness (and all of these are interrelated), other firms, which may be either less experienced or less knowledgeable about the product and technology, will seek new ways to offer higher limits and improved performance. Myopia stems from current product/service suppliers being "product provincial," i.e., not taking the lead in promoting the "creative destruction" of their product. Thus, the S-shaped product diffusion curve can be viewed as a "myopic curve" when change comes as a surprise or the firm fails to recognize the maturity of a product. Alternatively, it may be viewed as an opportunity curve if it prompts the initiation of strategic marketing efforts that either introduce new products and services or investigate ways to extend the life of products, even though their limits have been reached.

HOW TO AVOID MYOPIA: THE NEED FOR STRATEGIC VISION

According to Levitt and other writers, there needs to be a "strategic vision" of how customer needs should be satisfied.[1,2] This vision is best supplied by starting with the customer. The way to overcome myopia is to listen to what the customer wants and needs. It is vital for an industry to be viewed as "a customer satisfying process, not a goods producing process."[1] Proctor and Gamble (P&G) has been recognized for its heavy emphasis on consumer research as the basis for product design and marketing decisions. Many of the "excellent" companies profiled in In Search of Excellence had a "fanatical" viewpoint about identifying and responding to customer needs.[5]

According to Levitt, "the chief executive himself has the inescapable responsibility for creating this environment, this viewpoint, this attitude, this aspiration." Management "must push this idea (and everything it means and requires) into every nook and cranny of the organization. It has to do this continuously and with the kind of flair that excites and stimulates the people in it."[1]

Thus, the source of myopia is internal and the solution is mostly external: the customer. As Buck Rodgers, former vice president of marketing at IBM, put it: "the customer is perhaps your best source of information for long-term planning. After all, to survive, you must understand the customer's problems so you can provide solutions that withstand hard scrutiny over a period of time."[6] IBM relies on a number of customer input methods including personal interviews, survey research, the creation of specific industry councils (which represent top industry leaders), user conferences and training, etc., to understand the customer's needs and problems.

The managers of electric utilities today control the future destiny of their industry in terms of determining how loyal their customers are, who their competitors are, and how long their product will be desired.

WHAT CAN MANAGEMENT DO TO CURB MYOPIA? START BY LOOKING INTERNALLY

The prescription for overcoming myopia includes investigation of the following:

- Increased customer research and dialogue
- The structure of the utility
- The communication and decision-making processes
- The culture of the utility.

Increased Customer Research and Dialogue

Just as corporate and financial plans are routinely developed at electric utilities, so should there be an annual plan for investigating customer needs and benefits. Customer research should be of two types: qualitative and quantitative. Examples of qualitative research include focus groups and in-depth interviews, which are useful for identifying program and communication concepts as well as designing surveys. A series of customer encounters involving representatives of all departments and job levels should be planned. It is very important for even the "supply-siders" in a utility to have some contact with customers in order to better understand their preferences. There are a number of quantitative research methods that may be applied, including mail and telephone surveys. The results of this research should be widely disseminated within the utility so that a customer-oriented mentality emerges.

The Structure of the Utility

The structural organization of the electric utility may be a major inhibiting factor. Structure affects communication and the time required for decisions to be made. Most electric utilities are still organized to reflect the production orientation that exists. Information on customer wants and needs seldom reaches the generation and system planners who make most of tomorrow's marketing strategy decisions. Rates continue to be viewed as serving a revenue function and not a marketing function. Rather than being a market-driven entity in which all facets of the utility feel market pressure, it is more common for electric utilities to consist of specialized spheres of interest with no overriding reason to start with the customer. Its own internal bureaucracy often isolates a utility from its customers; those in the boardroom and in the general office, where most planning takes place, are sometimes the most isolated. This narrow focus is shared right on down to the district operations level, where transmission, distribution, and new business are typically given priority over responding to customer demand-side management (DSM) needs. This writer has seen utility executives stifled and government and customer directives ignored because of the insular nature of electric utilities. This insularity and the hierarchical structure of utilities cause them to move very slowly and to be reactive to changes in the market. Because many products of today are moving more swiftly through the product life cycle, the capability to respond swiftly to market changes becomes paramount. Without this capability, electric utilities end up marketing their services according to someone else's game plan. Close ties with the customer help prevent delayed responses to changing market trends.

The Communication and Decision-Making Processes

The communication and decision-making processes at many electric utilities are hierarchical rather than lateral. Today's marketplace requires much more interdepartmental communication. There is a need to institutionalize broad-based input to utility decision-making. Hierarchical structures need to be collapsed and more developed forms of lateral communication need to be developed (especially where rapid market changes are occurring). According to Warren Bennis, the leadership of any organization has fundamental responsibilities in situations where change is occurring. These responsibilities include:

- Directing and assembling human and physical resources to fit the required tasks given the new requirements that an organization faces.

- Building a collaborative climate. The synergy of different organizational units is more important than hierarchy.

- Facilitating adaptation -- considering business as usual as a one-way ticket to obsolescence.

- Creating organizational identity and building commitments.

- Acknowledging innovation and change as a means of continuing the revitalization of an industry.[7]

The changing utility customer marketplace demands these changes. The firms posing the greatest threat to electric utilities are already using this form of internal communication. Most of the snafus electric utilities encounter in developing strategic marketing plans relate to communication and authority problems which stem from both structure and process. As Rodgers noted, "an organization must constantly communicate its beliefs to its people,"[6] in order to incorporate these beliefs into every facet of the organization and ensure that they are reflected in tactics, strategies, etc.

The Culture of the Utility

Prior research by the Electric Power Research Institute (EPRI) has found that utility philosophy is an important determinant of the type of marketing program decisions that are made.[8] Other references on business management place increasing emphasis on the importance of organizational culture. Moreover, current organizational culture may be the first target that a senior officer needs to attack in moving from a production-based to market-based orientation. Corporate culture basically refers to the recurring patterns of behavior and belief transmitted across succeeding generations of management and employees.

Most firms that have been acknowledged in recent years as successful market-driven companies were noted for their strong corporate culture. In a limited survey of public and private organizations completed by McKinsey & Company, firms with uniformly outstanding performance had strong cultures.[9]

The strength of the culture of an organization typically shows itself in the lowest levels of an organization and in different functional departments. The following statements have been used by firms to help communicate business priorities:

- General Electric: "Progress is our most important product"; "We bring good things to life"
- IBM: "IBM means service"
- Sears, Roebuck: "Quality at a good price"
- Rouse Company: "Create the best environment for people"
- K-Mart: "Satisfaction guaranteed"
- Federal Express: "When it absolutely, positively has to get there"
- Proctor and Gamble: "Do what is right"; "Make it happen"
- McDonald's: "Quality, Service, Cleanliness and Value."

These "cultural" statements may have limited significance to us, but what is important is that, among employees at these firms, these statements are understood and internalized.

Having a strong corporate culture at utilities is important because it provides cues as to what employees should pay attention to and how employees can feel better about what they do. It could be hypothesized that a major barrier electric utilities face in communicating with stockholders and employees is that the prevailing culture points to the traditional product of generation or electric supply, as opposed to the provision of diversified energy services on both the supply and demand sides.

A key determinant of utility success in strategic marketing will be how prevailing utility culture reinforces strategic marketing and the role each employee has to play. The tools required to bring this about include senior management actions, not just words, and yet there is no doubt that senior management will also need help. This is why the careful placement of change agents and the use of symbols and rituals are required.[8] When it is felt that cultural change is necessary, a major part of senior management's time must be devoted to producing the symbols and rituals necessary to bring about a change. Effective responses to a market-based perspective will require a change in corporate culture.

Culture change is extremely important when the market environment is undergoing fundamental change, when an industry becomes more competitive, and when a company is getting larger or growing at a faster pace. The process that is designed to bring about this change must recognize the importance of consensus, trust, and the development of new skills through training, and must introduce employees to the changing environment around them. Organizations with strong corporate cultures usually have rigorous management training programs, e.g., McDonald's Hamburger University and IBM's training program. The implication is that in order for electric utilities to succeed in conducting the customer-oriented planning that will be required in the future, a systematic means of instilling this belief within a utility must be developed and implemented.

CONCLUSIONS

Marketing myopia stems from an inner-directed product focus which is based on a number of myths including economies of scale and unlimited product demand. The only way to overcome this nearsightedness is to develop a dialogue with customers and to invest in market research. Both environmental trend analysis and cross-sectional research will be required. A number of barriers inhibit the transition to a market orientation, including the structure, process, and culture of the utility itself. Senior management must possess and communicate the vision necessary to change these important elements in a fashion which parallels change in the external environment.

REFERENCES

[1]Levitt, Theodore. "Marketing Myopia." Harvard Business Review, July-August 1960.

[2]Hickman, Craig R. and Michael A. Silva. Creating Excellence: Managing Corporate Culture, Strategy and Change in the New Age. New York: NAL Books, 1984.

[3]Toffler, Alvin. The Adaptive Corporation. New York: McGraw Hill, 1985.

[4]Foster, Richard D. Innovation: The Attacker's Advantage. New York: Summit Books, 1986.

[5]Peters, Thomas J. and Robert H. Waterman Jr. In Search of Excellence: Lessons from America's Best-Run Companies. New York: Harper and Row, 1982.

[6]Rodgers, Buck. The IBM Way. New York: Harper and Row, 1986.

[7]Bennis, Warren G. and Philip E. Slater. The Temporary Society. New York: Harper Colophon Books, 1968.

[8]Synergic Resources Corporation. Marketing Demand-Side Programs to Improve Load Factor. EPRI EA-4267, Electric Power Research Institute, Palo Alto, California, 1985.

[9]Deal, Terrance E. and Allen A. Kennedy. Corporate Cultures: The Rites and Rituals of Corporate Life. Reading, Massachusetts: Addison-Wesley Publishing Company, 1982.

INDEX

O

P

R